BMW M3

JOHN DENNY

AMBERLEY

Acknowledgements

Thanks to my long-suffering wife, Sam. Without her support, my M3 experiences would have been limited to reading about them in books like this, rather than owning and enjoying them personally. To BMW Press Group Global (Chris) and BMW Group Classic (Ruth) for the images that fill in the blanks in my ever-growing collection. Thanks also to Munich Legends and FastClassics for some rarer vehicle images. And finally to my best mate, Phil, may he rest in peace.

The performance salon yardstick, often imitated, never duplicated.

E46 M3 CSL.

First published 2024

Amberley Publishing
The Hill, Stroud,
Gloucestershire, GL5 4EP

www.amberley-books.com

ISBN: 978 1 3981 1805 8 (print)
ISBN: 978 1 3981 1806 5 (ebook)

British Library Cataloguing in Publication Data.
A catalogue record for this book is available from the British Library.

Typeset in 10pt on 13pt Celeste.
Typesetting by SJmagic DESIGN SERVICES, India.
Printed in Great Britain.

Contents

	Introduction	4
1	The E30 BMW M3 – A DTM Legend	8
2	The E30 BMW M3 Special Editions	16
3	The E36 BMW M3	27
4	The E36 BMW M3 Special Editions	36
5	The E46 BMW M3	41
6	The E46 BMW M3 CSL	50
7	The E46 BMW M3 Went Racing	55
8	The E92, E90 and E93 BMW M3	58
9	The E92, E90 and E93 BMW M3 Special Editions	68
10	The F80 BMW M3 and F82 BMW M4	81
11	The F80 BMW M3 UK Special Editions	89

Introduction

In 1985 BMW turned their small practical E30 saloon car into a homologation special so that they could go racing. Built to satisfy Group A touring car rules, 5,000 road-legal copies of the racing car had to be sold within twelve months to qualify the vehicle to be used for racing.

Who would have thought well over thirty years later that the M3 name would have become such an icon?

In an era of 'race on Sunday, sell on Monday', never was it truer than in the case of the E30 M3, winning races in its debut season and continuing to win virtually every major

The BMW E30 M3 DTM racing car. (BMW Group Global)

event it ever competed in, becoming the most successful touring car of all time, only hanging up its boots in 1992.

This racing success helped to sell close to 18,000 vehicles, and we were hooked. With each new evolution came more features, more models, more technology and of course more 'M' power.

The first M3 was only available for the European and US markets, meaning it would be produced and sold as a left-hand drive. But the next iteration, the E36 M3, would be available to those of us who prefer the steering wheel to be on the right-hand side of the vehicle, and more importantly the gear stick to be next to our left hand.

Not only were we treated to a right-hand-drive car, but convertible and four-door models were also added to the range, so you could go racing around with your roof down, or convince the other half that you were not having a mid-life crisis, but you were merely buying a practical family car that just happened to look a bit like the one from the telly...

From that point on it's fair to say that the M3 brand took off, becoming not only BMW's flagship 3 Series model, but also the sports saloon gold standard.

As a result, the second generation E36 M3 would see a host of improvements as well as considerable growth. Although not designed specifically for racing, motorsport DNA is still evident in the development of the cars and has helped with enhancements in performance and driver aids, with not only engineering know-how, but software and computing power too. Though aimed at the executive market, that wouldn't stop the M Division from sprinkling the car with M suspension, M sports seats and most importantly M power in the form of the latest iteration of the M engine, now in a familiar six-cylinder format, with a healthy increase in both size, power and torque. It wouldn't be long until the E36 M3 would find its way back onto the racetrack, and in 1996 the M3 would compete in the GTS-2 class and then re-designated GTS-3 class, and start collecting wins and titles for Manufacturer, Team and Driver categories.

M purists would be treated to something quite special in the next iteration, as flared wings, wide arches and a front end that says 'move over' would return with the E46 M3, and though the size and weight would still be on the up, so was the power. To get all the power to the road the E46 M3 would receive one of the best chassis built to date by BMW, and when it came time to roll out the all-important special edition, BMW's M Division would create one of the definitive drivers' cars of a generation in the form of the E46 M3 CSL, a car that can certainly live up to the name of being 'the ultimate driving machine' – even if Mr Clarkson would suggest it had a carboard floor.

The fourth generation E92 M3 would be something of a culture shock, as for the first time ever BMW would use a new lighter, more powerful mass-produced 4.0 V8 in the M3 that could deliver over 400 hp thanks to engine controls that would not be out of place in a spaceship. The four-door would be returned to the line-up alongside the coupe and a hard-top convertible, and technology would start to feature heavily on all the models, with electronically controlled dampers (EDC) returning, an Active M differential, electrically adjustable steering, and for the first time the offering of a dual-clutch seven-speed gearbox that would feature eleven electronically controlled drive programs. And if that wasn't enough, you could store your favourite set-up to be recalled at a touch of the new M button on the steering wheel.

E92 M3 M Sport steering wheel with 'M' button.

The E9x M3s would start to incorporate BMW's 'EfficientDynamics', aimed to keep fuel consumption and CO2 emissions as low as possible while improving dynamics and driving experience. This ethos would see the E9x M3s get a brake regeneration system, as well as auto start/stop technology for the first time in a model with more than four cylinders.

E92 M3 with lightweight carbon fibre roof.

F80 M3 S55 engine with CFRP precision strut.

Light weighting would become high on the design priorities, evidenced by the use of more lightweight materials such as aluminium in the body and suspension components, and a carbon-reinforced plastic (CRFP) roof being adopted as a design cue from the E46 M3 CSL, designed to reduce weight and lower the car's centre of gravity.

As we become more aware of our carbon footprint, BMW, as with all manufacturers, has been forced to make their vehicles more environmentally friendly, so the fifth generation F80 M3 would not receive the high-revving lightweight V8 that we saw in the fourth generation M3 but a new smaller 3.0-litre unit. But it wasn't all bad, as not only did we get more miles to the gallon, but a couple of turbochargers, more horsepower and a shed load of torque! Light weighting and EfficientDynamics would still feature heavily in the fifth generation M3, showing that the M Division was still focused on the driving experience as the overall priority.

Now in its sixth generation, though a far cry from its grandparents, the G80 M3 is still delivering the basic M3 package. A practical four-door saloon car that can be a weekend warrior or handle the daily commute to the shops, it can safely transport the kids to school or tear up the racetrack, and it's still the motoring journalists' drift car of choice.

1

The E30 BMW M3 – A DTM Legend

'Let's Go Racing!'

BMW wanted to enter the German Touring Car Championship, so they took an E30 3 Series, dropped in a high-revving 2.3-litre inline-4 engine with 197 hp, flared arches and a tuned chassis, built 5,000 units to satisfy homologation rules, and off they went, collecting several championship wins in its first year.

On top of that came countless victories in the world and European championships, as well as twenty-four-hour races at the Nürburgring and Spa-Francorchamps. As with the BMW 635 CSi, the success story began immediately with victory in the first race at Hockenheim in 1987. Harald Grohs was the man on the top step of the podium again, but once more he had to play second fiddle to a fellow BMW driver at the end of the season; another man who was crowned champion without winning a single race. This time it was Eric van de Poele as lightning struck for the second time.

Each year BMW drivers were among the main contenders. 1989 delivered another title win, this time for Roberto Ravaglia, and it was Ravaglia who won both races at the final round of the 1992 season at Hockenheim to bring the first chapter of BMW's DTM participation to a close with a bang.

When the first car rolled off the production line BMW could not have dreamed of the success that would lie ahead. Competing in every permissible series up to world championships, the E30 M3 would go on to win over forty-one races, making it the most successful touring car of all time, a title that is still true some thirty years later.

BMW would go on produce over 18,000 road cars between 1986 to 1991, adding a couple of Evolutions and a couple of racing-inspired limited editions before the curtain would fall on this sports car icon.

In the creation of the M3, the donor car would share no other panels with the M3 bar its bonnet. The car featured flared front and rear wings, a re-profiled rear windscreen, raised boot lid and an enormous rear spoiler, along with other M3-specific features including a deep front spoiler, with ducts to direct cool air to the brakes and over the oil cooler, extended side sills and special bumpers.

Above: Eric van de Poele with his E30 M3 DTM, Berlin, 1987. (BMW Group Global)

Right: E30 M3 with its extended side sills and special bumpers. (BMW Group Classic)

M3-specific body parts were all fabricated from SMC composite. The front and rear windscreens were bonded in place to improve rigidity, and the chassis was updated to include a quicker steering rack (19.6:1 to the standard car's 20.5:1), and three times more castor was added, along with stronger wheel bearings, wider tracks, and revised shocks with shorter, stiffer springs. The rear anti-roll bar was increased to 19 mm and the front anti-roll bar was linked to the front struts.

Inside, the E30 M3 shares its basic architecture with the standard E30 3 Series. However, all M3s are equipped with manual sports seats that were available in leather, cloth or a combination of the two; rear seats were formed to fit two passengers. A black headliner and an M Specialist dash featuring red needles and the all-important M logo, together with an oil temp gauge in place of the economy meter, would finish off the special cabin.

Cars without a driver-side airbag have a leather-wrapped M three-spoke steering wheel in one of two designs: the M-Technic I with a slimmer hub and the M tri-colour stripe on the centre spoke, used on M3s built before September 1989. After this date, the M-Technic II steering wheel with a thicker rim and the M logo on the central spoke was used.

The gear stick, finished in leather, includes an M stripe with the 'dogleg' shift pattern. The parking brake handle boot is unique to the M3 and is also trimmed in leather.

E30 M3 with full-leather sports trim. (BMW Group Global)

E30 M3 'M' Specialist dash featuring red needles and the all-important M logo. (BMW Group Classic, Robert Kröschel)

The M3 would also receive a BMW check control vehicle monitor, which was positioned on the headlining above the rear-view mirror.

The elephant in the room must be addressed here, and that is the steering wheel location, as it was on the wrong side of the car, the left, as the E30 M3 was never produced for the UK market, which was a real shame, as though there are plenty of imported examples on our shores, most, if not all, are still left-hand drive.

Some UK dealers converted a few cars to right-hand drive, but changes under the bonnet to accommodate the steering affected performance as modifications had to be made to the exhaust manifold and this didn't appeal to the enthusiasts but as most of the world does seem to prefer driving on the left-hand side of the road, having a steering wheel on the left does make sense.

The aggressive stance was compounded by the M3 E30 being lower than the standard car, and handling was further improved with independent front suspension with McPherson structs, trailing arms at the rear and larger anti-roll bars all around. Stopping was applied with floating front callipers, 280-mm vented discs and ABS.

Differences from the standard E30 were not limited to extensive body modifications; they also included five-stud wheel hubs paired with lower-profile 15-inch tyres running

Improved E30 M3 suspension with independent front suspension with McPherson struts. (BMW Group Classic, Robert Kröschel)

205/55/15, which would continue to increase in size to 16 × 7.5 with 225/45/16 tyres on later sport and special editions.

Moving to the things you don't see right away but you can definitely hear, the new high-revving 2.3-litre four-cylinder double overhead cam produced a healthy 197 hp (200 PS) without any emission controls, designated the S14B23, easily identifiable by the

BMW M power signature raised lettering on the cylinder head, which would be a common theme throughout BMW's special M engines.

Based on the M10 cylinder block, an engine that was already proven in motorsport, the S14 would contain a five-fold mounted crankshaft designed to be insensitive to torsional and bending vibrations, as is common in motorsports at very high speeds due to their short length, and would be mounted to a dogleg Getrag five-speed, keeping all the main gears in an H pattern for quicker gear changes.

The engine would be controlled by Digital Motor Electronics (DME), which automatically adjusted the cold start behaviour without a cold start valve and, depending on the altitude, could adapt the ignition angle and/or mixture formation. The engine relied on two separate throttle bodies, each incorporating two throttle butterfly plates. The cylinder head would receive some attention with the inlet value bore increased to 37 mm with 18 degrees of incline, and the exhaust bore increased to 32 mm with a 20-degree incline, resulting in a near-perfect V shape to improve charge response.

Looking back to the past, we find that EDC, now an option in the M3 from E9X onwards, was first made available to the E30 M3, with its three driving modes, with the sportiest of the three best suited for the racetrack. But sadly as the cars got older, failed units would be switched out for standard units, meaning should you find one with the EDC intact and working, it's quite a gem.

The legendary high-revving E30 M3 engine (S14). (BMW Group Classic, Robert Kröschel)

E30 M3 side by side with its racing counterpart the E30 M3 group A. (BMW Group Classic, Robert Kröschel)

Considering where the M3 had come from, with its world champion-winning DNA, it's hard to believe that you would want to chop the roof of one of the best handling chassis that BMW M Division had created at this point, but the marketing department must have had some say, as 786 convertibles would be produced during the E30 M3's production run.

The popularity of classic sports cars with a racing heritage has never been higher, and with that popularity comes high demand. As a result, the E30 M3 prices have skyrocketed. Used E30 M3s can now be seen trading hands for over £40,000, and you don't see many for sale, especially good rust-free examples.

Older BMWs will be the first to suffer parts becoming obsolete, and with the E30 M3 having so many bespoke parts, and being a lower-production run car, it could lead to parts being scarce. Rust on the E30 M3 is an issue, and of course, the electronics get less reliable the older they get. Worn trim is an issue as replacement cloth again is getting harder to come by.

In general, if you are lucky enough to be looking at E30 M3 ownership, you'll be most likely purchasing from a fellow M3 enthusiast, so the car will hopefully have been well maintained and usually garaged. Just check all the service history, and check for accident damage, poor repairs, or emerging rust, and if you are unsure, invest in an RAC- or AA-type inspection, as it's well worth the money to protect yourself. And of course, the inspector will not be all wide-eyed looking at the potential ownership of one of the more recognised M cars of all time.

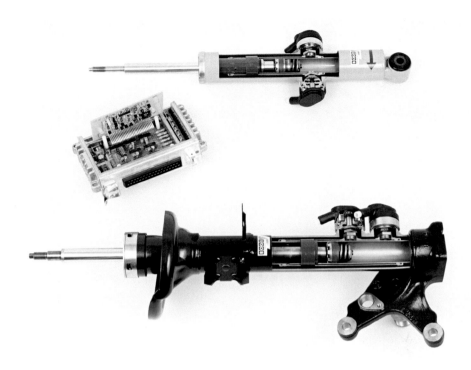

E30 M3 variable chassis and suspension tuning electronic damper control (EDC). (BMW Group Classic, Robert Kröschel)

E30 M3 now ready for the summer. (BMW Group Classic)

2

The E30 BMW M3 Special Editions

1987 E30 M3 Evolution 1

BMW didn't stand still with the E30, and as time went on, so did the progression of the DTM, resulting in changes to the rules and regulations. BWW responded by giving the M3 some mechanical and aerodynamic updates. To achieve this, BMW Motorsport built a lightly modified E30 M3, the 'Evolution 1'. To satisfy the homologation rules 500 cars had to be built, and as with the first M3, this would have to be within a year.

Production took place between March and April of 1987, to align with entering the cars for races in the same year. Visually there were no changes to the exterior of the M3, and you would be hard-pressed to tell the difference inside. Manual Recaro sports seats were

E30 M3 – the first Evolution. (BMW Group Classic, Robert Kröschel)

upholstered with houndstooth check cloth centres, plain grey fabric bolsters and grey PVC backs. The rear seats would again be individually formed for the comfort of your two amble-sized passengers.

The standard E30 dash and instrument binnacle was reused, though had a set of new red-needled VDO instruments, with the addition of a 160 mph (260 kph) speedometer and an 8,000 rpm rev counter. The large dials were split by a BMW Motorsport logo and again the oil temperature gauge replaced the economy meter, as seen in the lower models. A special three-spoke leather-rimmed M-Tec I steering wheel was also standard, along with a Motorsport-branded gear lever, tinted glass and electric door mirrors.

1998 E30 M3 Evolution II

The next Evolution would be the Evolution II, produced between March and June of 1998. Again another 500 units were built by BMW Motorsport.

This time, the homologation special received major changes to the standard M3. The engine was built with high-compression pistons (10.5 to 11.0:1) and the air intake was reworked – a 265-degree intake cam now featured along with a lightened flywheel and engine management. This saw a power increase to around 220 hp with 181 lb-ft of torque and became the most powerful 2.3 S14 engine that BMW Motorsport ever produced that would be used in their road-going M3s. BMW was also able to gain significant weight loss by using an aluminium head, which improved handling over the car's competitors, unmistakable by its tri-colour-painted rocker cover.

E30 M3 Evolution II with a deeper front spoiler, fog lights replaced by brake ducts. (www. fast-classics.com)

Above: E30 M3 Evolution II tri-colour-painted rocker cover. (www.fast-classics.com)

Below: E30 M3 Evolution II rear with lip spoiler. (www.fast-classics.com)

E30 M3 Evolution II Berkaro cloth with silver bison leather bolsters. (www.fast-classics.com)

To go with the extra power, BMW Motorsport continued its weight-saving regime, cutting loose an additional 10 kgs through the addition of thinner window glass and lighter bumpers and boot. The exterior received some aero improvements too, with a deeper front spoiler, brake ducts where the fog lights used to be, and a lip spoiler on the boot.

The Evolution II would feature the previously used five-speed dogleg gearbox, set up to provide a quick change between second and third gear, utilising the same ratios but a shorter final drive, shortened from 3.25:1 to 3.15:1, and would sit pretty on a nice set of 16-inch BBS cross-spoke alloys.

Only available in Misano Red, Macao Blue and Nogaro Silver, the Evolution II had only one interior choice: Silver Uberkaro cloth with silver bison leather bolsters.

The interior featured the tri-colour M lapels on both front and rear seats, with M3-embossed sill covers. Rounding it all off was a special dashboard plaque with the designated build number.

1988 BMW E30 M3 Europameister

To mark the continued racing success of the M3 and its drivers, including Roberto Ravaglia who would go on to get his own special, BMW made a short run of 148 Europameister special edition celebration models.

E30 M3 Europameister deeper front spoiler with fog lights. (www.munichlegends.co.uk)

The run would only last around three months at the end of 1998. Finished in Macao Blue metallic, they would get a silver extended Nappa leather interior, special edition door cards, the obligatory M stripe monogrammed seats, as well as the all-important series plaque signed by Roberto himself; everything else was mainly as the standard M3, including the world-beating S14 4 pot motor.

1989 E30 3 Series M3 Johnny Cecotto

To celebrate the DTM success of their Venezuelan driver Johnny Cecotto, BMW released another special edition. They featured a few special add-ons, such as M3 Evolution II front and rear spoilers, thinner glass on the rear window, chrome tailpipes, and 7.5×16-inch alloy wheels with one-off metallic black centres.

Moving to the inside, the standard seats were unholstered with a special Anthracite or Silver M-Technic cloth with bison leather side bolsters.

The all-important plaque with the signature of Johnny Cecotto, plus the exclusive number of each car, would feature on the centre console, showing the occupants they were in something a bit special, and if you were to peek under the bonnet you would be greeted by a further reminder of this model's uniqueness with its individual M power S14 engine, topped with a

BMW M Power

Above: E30 M3 Cecotto 'BMW M power' poster. (BMW Group Classic, Kirchbauer)

Below: E30 M3 Cecotto sports seats with special Silver M-Technic cloth with bison leather side bolsters. (BMW Group Classic, Robert Kröschel)

E30 M3 Cecotto S14 Engine – with its striking red rocker cover. (BMW Group Classic, Robert Kröschel)

bright-red rocker cover that was uprated and now fitted with a catalytic convertor that would go on to be the standard engine for the European specification M3s from September 1989.

1989 BMW E30 M3 Ravaglia – Rarest of the Rare

When it comes to car collections, there's rare and there's really rare. The E30 M3 Roberto Ravaglia Edition is possibly the rarest M3 of them all. Based on the Evolution II, BMW produced twenty-five examples for the UK market, though still with that steering wheel firmly mounted in the wrong place...

Together with Johnny Cecotto, Roberto Ravaglia elevated the E30 M3 to legendary status, winning pretty much every race it turned up to. The cars were available in Misano Red or Nogaro Silver, and sat on black 16-inch BBS alloys with contrasting silver rims. The engine would be adorned with a body-coloured rocker cover, and the interior was similar to the Johnny Cecotto models, with M-Technic cloth with bison leather bolsters and would again receive the all-import special edition plaque.

1989 E30 M3 Ravaglia special edition in Misano Red. (www.munichlegends.co.uk)

E30 M3 Ravaglia special edition numbered plaque with the driver's signature. (www. munichlegends.co.uk)

1990 BMW (E30) M3 Sport Evolution

As the E30 M3 neared its end, BMW would finish the run with 600 final edition specials, the Sport Evolution. The most noteworthy addition would be found under the bonnet, where the 2.3 S14 (B23) motor would see an increase in the bore from 84 mm to 95 mm, combined with a long stroke crank capacity that would be increased to 2,467 from the previous 2,305. More treats were given to the engine, such as larger valves, a more aggressive camshaft and a special oil piston cooling system. The net result was the S14 (B25), with an increase of power to 238 hp that could be enjoyed all the way to the dizzying height of 7000 RPM and would see torque peak at 177 lb-ft. Red plug leads were a new addition, letting all who looked on know that this was no ordinary power plant.

As with the Evolution II, the car would receive the same lightweight bumpers and boot lid, as well as thinner window glass and a smaller fuel tank. Cooling was further improved with changes to the grille, to go with the same fog light removal treatment that was seen before on the Evolution II.

Internally the car would receive the Sport Evolution trim, red seatbelts and the all-important Sport Evolution plaque.

The most powerful version of the S14 fitted to the E30 M3 'Sport Evolution' engine. (BMW Group Classic)

1989 BMW E30 M3 Sport Evolution. (BMW Group Classic)

With homologation ever present, the Sport Evolution would see some exterior changes, such as wider and taller front wings to make more room for larger wheels when racing. Uprating the anti-roll bars, increasing the steering caster angle, upgrading the wheel bearings, adding stiffer springs and uprating the shock absorbers all added to the car's handling. These changes also saw the cam sit 10 mm lower than the standard E30 M3.

The most noticeable feature of the Evolution Sport was the adjustable rear wing, with three positions to choose from – Monza, Normal and Nürburgring – paired with an adjustable front spoiler lip. It was something quite special.

Left: E30
M3 Sport
Evolution
interior with
its distinctive
red seatbelts.
(BMW Group
Classic)

Below left:
E30 M3 Sport
Evolution
front spoiler
standard
position.
(BMW Group
Classic)

Below right:
E30 M3 Sport
Evolution
rear spoiler
standard
position.
(BMW Group
Classic)

3

The E36 BMW M3

1994 BMW E36 M3 3.0

When the second generation M3 was released to the world in 1992, it felt like the M3 had grown up; it was now a larger heavier proposition. In coupe form the new E36 model tipped the scales at 1,460 kgs – 295 kgs more than its lighter sibling – and the growth didn't stop there. The wheelbase had also been extended by 130 mm to 2,700 mm, with an overall length of 4,433 mm, an increase of 108 mm. Front and rear axles received more of

E36 M3 3.0 Coupe in Avus Blue Metallic with Evolution front splitter.

the same, with an increase in the front axle width by 15 mm to 1,422 mm, and an increase of 23 mm in the rear to 1,438 mm. To cover this increased track, the M-tuned chassis had to grow a further 65 mm to 1,710 mm in width.

It wasn't all bad though, as BMW had obviously thought about all this midlife spread. They added a couple of cylinders under the bonnet, taking the power plant to 3.0 litres delivering 286 hp, with a respectable 320 Nm/236 lb-ft torque, up from 197 hp and 240 Nm/ 177 lb-ft torque, assuming you were in a UK or Euro specification car – the US spec S50 would closely resemble the normal production M50 engine and produce less power (around 240 hp).

The new engine designated S50 was an iron block six-cylinder twenty-four-value twin cam with an aluminium head. Along with its extra cylinders, it also inherited a single VANOS unit from the M50 engine, helping to propel the E36 M3 to 62 mph in six seconds.

VANOS: In technical speak 'The BMW VANOS (variable *nockenwellen steuerung* in German) is a variable valve timing technology developed by BMW. The system variably adjusts the timing of the valves by changing the position of the camshaft relative to the drive gear.'

In layman's terms, it is BMW's V-TEC or VVTI. It's a clever method of giving your car the best of both worlds: low-down torque when you need it for pulling off, and then high-end power when you are opening the taps.

At this point, it is fair to say the internet is awash of VANOS horror stories and honestly, there is a lot of truth to the fact that the VANOS can leak oil, the solenoids can fail, or worse

E36 M3 3.0 S50 engine with single VANOs and 24 Value DOHC.

E36 M3 3.0 in classic Dakar Yellow.

case you get a metallic rumble as the inner gears rattle about like marbles in a bucket. The issues are a result of what happens when you have complicated timing systems that have not been maintained within their factory specifications, which sadly is often the case when the cars have piled on their miles and years, then found themselves in the hands of owners who didn't or couldn't afford to complete a detailed maintenance regime.

In short, they fail, and they failed a lot, so much so that this engine got quite a bad reputation, and what compounded the issue was that after BMW had stopped fixing them under warranty, the normal cost for a full replacement unit and fitting was well over £2,000. At the low point of trading, some E36 M3s were sold for as little as £4,000, making fixing a broken vehicle a rather scary proposition.

If you were lucky you could get away with an oil seal or a new solenoid, but in most cases, and especially if the dealer was doing the job, it would be a pricy affair. However, it must be said that when properly maintained the VANOS system is a very clever piece of kit and makes for an exhilarating driving experience when you are pushing hard.

Benefiting from its German engineering, a BMW E36 M3 with a good service history is a safe proposition on the second-hand market, even with the VANOS horror stories, but as the cars are getting older, they start to develop rust in the front and rear arches if they haven't received regular attention. The boot can also crack around the boot lock mechanism as a result of one of the mounting points failing, allowing the lock to flex, and this is worse on a car with the M spoiler. Along with the dreaded rust, and general lack of maintenance that is often evident on older, cheaper, higher-mileage examples, one common failure point is the rear shock top mounts, but they are a cheap item and easy to fix.

When purchasing an older E36 M3, check the exhaust and the catalytic converters are in good order as they can rust, especially the back box and around the catalytic converter's connection points. Door rubbers don't seem to stand up well to the test of time, and they are getting quite expensive now from BMW, so check they are not hanging off or ripped. Most external trim that's looking worse for wear can be still purchased from the BMW parts department.

The E36 M3 has a flexible hydraulic clutch pipe that can expand, meaning the clutch can drag; a new pipe, or a branded updated pipe from eBay or other websites, is often a cheap fix and a good improvement. This all stacks up alongside the usual suspects, such as the clutch's condition, lambda sensors and bushes wearing out, specifically the front lower arm bushes and the rear trailing arm bushes.

Following a DTM legend was always going to be a tough ask, and it's clear that from a technical standpoint with revised modern suspension, brakes and the improved chassis with its new wider track, there was plenty to boast about. The only thing that perhaps was a bit flat was the styling, even if there were now some M specific colours added to the E36 range, such as Daytona Violet and the ever-popular Dakar Yellow.

The E30 M3 was unmistakable with its wide arches, prominent rear spoiler, bulging chin and rear quarter rake, not to mention in the UK, the steering wheel was on the wrong side, and was recognisable as a weekend racer that was missing its racing suit, shod with a pair of sensible shoes, and of course, a full complement of seats for its passengers.

E36 M3 'Vader' leather interior.

Though the E36 M3 came with all the practicality of a family-friendly four-door saloon to go with the coupe, the basic body shell was almost unrecognisable to the average Joe, so how would people would know you had shelled out the big bucks for the top-of-the-range Beemer? The E36 M3 could easily be confused with the lower-specification sports model, or in some cases, dare we say it, a 316i.

Above: E36 BMW M3 Evolution among good friends.

Right: The M3 Evolution brochure.

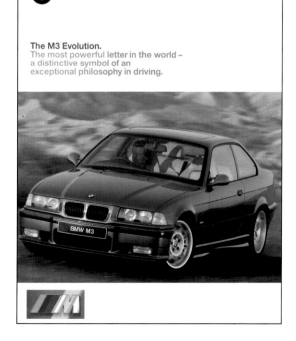

The M3 Evolution.
The most powerful letter in the world –
a distinctive symbol of an
exceptional philosophy in driving.

The E36 M3 was not designed primarily for competition use, but rather as a road-going GT car, therefore was more reserved and hid its potential well, so it could be sold and used as a daily driver; a car that had a good-sized boot, space to fit five people in relative comfort, and should you so desire it and were lucky enough to be on an autobahn, sprint to the top end of the speedo and pretty much sit there all day long without too much drama.

With the increase in wheel track, weight and power were managed by one of the most copied BMW wheels of all time, the Style 24 M Spoke Lightweight alloy with their BMW Motorsport credentials literally stamped into them, with size now increased to 17 inches, wearing 235/40/17 Michelin Pilot Sports all round if you opted for a square set-up, or 225/45/17 up front and a wider 245/70/17 at the rear, with the staggered set-up.

On the inside there were not many changes from the standard model; however, if you opted for the coupe model, you would be greeted by a 1970s throwback in the guise of the 'Vader' seats – if you know you know – unless you grabbed a convertible or four-door, as they had regular-style front seats, though offered in sumptuous leather, and the obligatory M logo that would appear below the speedo and rev counter.

The E36 M3 would get a set of new two-pronged electric-heated sports mirrors that have come to signify M3, as well as a shotgun stainless steel twin pipe, and the M3 logo on the boot and side M sport mouldings – but other than that, quite a subtle affair.

1996 BMW E36 M3 Evolution

In 1996 the M3 would receive its Evolution treatment. The styling was largely unchanged, and to the trained eye there was not much difference between the two models other than if you opted to have the front Evo Splitter, which again as before was still very discreet, and there were changes to the front grille that would see the BMW kidneys start to grow, a trend that would never be truer than it is today!

The M Division replaced the E36 M3 3.0 with the faster and more refined 3.2-litre version. For the European market this model was fitted with the new S50 B32 engine, an evolution of the S50 B30, which would now produce a claimed 321 hp thanks to the addition of a second VANOS unit and 200 more CCs, as the Evolution's new S50 B32 was now a 3.2-litre with an additional VANOS unit acting on the exhaust as well as the intake, as was the set-up of the previous engine. Sadly the VANOS would have similar issues to the Single VANOS, but now it was Twin unit, so double the issue. Fortunately there are specilists such as Mr Vanos (give him a Google) who has been rebuilding and upgrading VANOS units in the UK since 2009. The author has had a rebuilt Mr Vanos Unit in both a E36 Z3M and E46 M3 to date.

The revised engine was one of the first to produce over 100 hp per litre. Even if it was perhaps on an engine stand in a hermetically sealed engine room, it's still pretty impressive, and to help those extra ponys onto the tarmac, the five-speed ZF was replaced in favour of a six-speed manual gearbox as well as some detailed refinements to the chassis. For the US (and later Canadian) market, the M3 received its own engine, the S52, plus ASC+T traction control and a host of chassis upgrades. Both versions were also treated to a few cosmetic alterations.

The Evolution did however bring with it a rather surprising party trick! There you are in your M3, it says manual on the logbook, it has a six-speed manual gearbox, and it has a clutch and

1998 E36 M3 Evolution Coupe with no rear spoiler.

a slave cylinder. But 'look no hands', as you change gear as if it were an auto! Yes, you guessed it, it's the SMG, BMW's sequential manual gearbox, and if you thought that VANOS was clever, this is in a different league, though perhaps in retrospect it was a bit ahead of its time.

1998 would see both Ferrari and BMW race to launch the first F1-style sequential gearbox, and BMW got there first. Now let's be fair, it was an odd thing in automatic mode, which was a notion in an M3 and BMW's 'ultimate driving machine' no less. It was sometimes not sure what to do, as the second ECU controlled the gear changes and clutch control with a clever hydraulic system, and it had more sensors than were needed to land a lunar module. Driving around in this mode was a bit 'interesting', shall we say.

E36 M3 Evolution Convertible. (BMW Group Global)

This technology would be further advanced in the later E46 M3 model, and even the E60 M5 and E63 M6, but eventually an automated single clutch was considered too unreliable and was phased out in favour of the dual-clutch system. The SMG sadly became another famed frequent failure point, mainly thanks to the ever-increasing popularity of internet forums.

The author has had over twenty M3s in the last twenty years, and among the rank of M3s, five of them featured an SMG gearbox; one used to pop out of gear, one failed and was in such a state the car was converted to manual operation, another had its gear position unit fail, and another had the traditional salon pink relay failures. An 80 per cent failure rate – and that's not hearsay, that's pure experience.

It must be said, however, the SMG when working correctly, slotting through the gears with your foot buried in the carpet with no let-off whilst the computer riffles through the gears, is quite something, and in 1998 it must have felt pretty special, safe in the knowledge that if you broke it, BMW's warranty had you covered.

New wheels were added to the line-up with the M Contour II and forged M Double-Spoke, and the four-door got its own new design in the guise of the M Contour II, another variant on a BMW design house classic. The wheel design also incorporated extra safety as the tyre bead was secured to the rims, keeping the tyre safely in place in the event of a sudden flat.

Floating hubs were now added to the front braking system, helping to dissipate heat by allowing the unit to expand with minimal stresses when heated, therefore helping to reduce brake fade. BMW at this time was still producing M cars with single-piston floating callipers, which, though adequate for driving on UK roads with their front 315-mm vented discs, generally were not considered a real solution for track use, and were often replaced by big brake kits.

During the E36 M3's production run there would be four-door saloons and convertibles produced with both standard M and Evolution trim, though to get the full M3 goodness you would have to opt for the coupe. By the time the E36 M3 came to the end of its life

E36 M3 Evolution four-door saloon. (BMW Group Global)

a total of 71,242 cars would be produced, which was a huge increase on the E30's initial M3 run – split across the three platforms: 46,525 coupes, 12,114 convertibles and 12,603 – showing that the M Division had developed a winning formula.

The E36 M3 Goes Racing

In 1995 BMW once again picked up their 'race what they sell' baton, as it competed in the IMSA (International Motor Sports Association) stateside with the E36 M3 GTR, a wide-bodied variant of the M3, wearing a plethora of sponsors from Valvoline, Red Bull and Yokohama to name but a few, campaigning in the GTS-2 class and then re-designated GTS-3 class.

The car didn't initially dominate as the former model had, but after the next year's development, and changes in the IMSA homologation rules allowing for the higher output Euro Spec engine, and full seam welded shells coming from the factory, the E36 M3 GTR started picking up wins and went on to claim the Manufacturers' title in 1996, followed up in 1997 with three titles for that year: Manufacturer; Team; and Driver, thanks to their driver Bill Auberlen. BMW and Prototype Technology Group would dominate GT racing throughout 2006, amassing fourteen championships including wins at the biggest races in the US, including at 'Daytona' no less.

As the cars were coming to the end of their racing life as the E46 chassis was well on into its production, 1998 brought with it more success with GT2 race wins and BMW taking home other GT3 Manufacturers' titles in the USRRC (United States Road Racing Championship), BMWs first for the series. After claiming two Manufacturers' titles in a single season, the M3 took home the USRRC Team title thanks to (PTG) Prototype Technology Group.

PTG and BMW's effort went on to ensure that the second generation M3 racers had firmly written a chapter of racing's history.

E36 M3 GTR race car with the US spec E36 M3 (CSL) Lightweight. (BMW Group Global)

4

The E36 BMW M3
Special Editions

As with the E30, the E36 M3 would receive the special edition treatment, and in this iteration, thanks again to motorsports' influence, in 1995 BMW launched the M3 GT Coupe, a limited run of 356 cars finished in British Racing Green with matching two-tone leather and Alcantara interior, and a host of track-focused goodies and even a touch more power. But as with the E30, they were produced in LHD for the Europeans.

Advertising brochure for the European M3 GT Coupe. (BMW Group Global)

Even worse, a very limited run (126 units) of E36 M3 CSL Lightweights, weighing in at 1,338 kgs, some 177 kgs lighter than the standard coupe, was offered to the US market. It was finished in white, with M Sport checked flags draped across the bonnet edge, with the interior receiving some special weave carbon fibre accents and the all-important numbered plaque.

M3 GT Individual

Finally, we Brits would see a right-hand-drive racing-inspired special, in the guise of the M3 GT Individual.

Internally referred to by BMW as 'M3 mit 'GT'-Optik' (BF92), BMW produced a limited run of fifty cars for the UK market, produced in British Racing Green and based on the 3.0-litre M3, with an upgraded ECU, increasing power to 296 hp. The car featured an adjustable front splitter, rear GT Class II spoiler and a Motorsport strut brace, as with the GT Coupe.

The exterior was also treated to M3 Side Moulding Badges and a set of clear front indicators, whilst the interior would receive an Anthracite interior trim with the leather

Right: The M3 GT Individual in British Racing Green. (BMW Group Global, Barry Hayden)

Below: M3 GT Individual handbook excerpt detailing adjustable spoiler operation. (BMW Group Global)

M3 GT Coupé

Your M3 GT Coupé's advanced aerodynamics and sports suspension combine maximum road grip with outstanding handling. Tyre adhesion and stability are noticeably enhanced, and aerodynamic lift is considerably reduced.

For use on public roads, it is not necessary to make modifications to the front or rear spoiler, even if a dynamic driving style is adopted.

It is therefore possible to dispense with adjustments to the front spoiler and the addition of a rubber lip to the rear spoiler (though depending on the situation, this can be of value for motor racing).

Note:
If the front spoiler is extended or a rubber lip attached to the rear spoiler, the car's top speed is reduced and fuel consumption and tyre wear are increased.

36 95 01 252

Note:
Due to the special front spoiler, ground clearance is considerably restricted, especially with the spoiler extended. Drivers are therefore advised to take care when entering underground garages, driving over kerbs etc.

36 95 01 253

Note on automatic car washes
In certain types of automatic car wash, the blower has guide wheels which may not clear the rear spoiler in time. To avoid this, consult the car wash operator first so that damage can be prevented.

Above left: The M3 GT Individual bi-plane rear spoiler. (BMW Group Global, Barry Hayden)

Above right: M3 GT Individual interior – Anthracite and Mexico Green Nappa leather interior. (BMW Group Global, Barry Hayden)

Left: M3 GT Individual front spoiler with corner extensions and polished 'BMW Motorsport' double-spoke alloy wheel. (BMW Group Global, Barry Hayden)

M3 GT Individual engine with 287 hp and Motorsport strut brace. (BMW Group Global, Barry Hayden)

seat centre and door inserts, and grab handles in Mexico Green Nappa leather were also present, with seat side parts and headrests wrapped in Amaretta Anthracite and the all-important M stripes on backs of front and rear seats.

Sadly, we were not to receive the carbon fibre interior as with the GT Coupe, but got graphite bird's-eye maple wood inserts on the centre console and glovebox instead.

The GT Individual would sit on 17-inch polished forged M Double-Spoke alloy wheels, 7.5J front and 8.5J rear, with suspension like the standard M3 but with shorter stronger springs to take the extra downforce generated by the fully extended front splitter and rear spoiler. The car would be produced for homologation reasons – Internationale FIA-GT-Series, Division II, IMSA GT-Series USA and Internationale Langstreckenrennen.

M3 Imola Individual (GT2)

Then as a last hurrah, as the E36 M3 was coming to the end of its production run, BMW UK decided to create a run-out special, and for once this would be limited to fifty units for the UK market only, with right-hand drive based on the standard coupe with the styling cues from the GT Individual.

What they created was the Imola Individual, affectionately known as the GT2. Based on a manual 3.2-litre M3 Evo, the GT2s had the same interior stylings as the previous GT Coupes but would pair the Anthracite interior trim, seat side parts and headrests with the door insert, seat centres and grab handles in Red Nappa leather. Features such as Harman

E36 M3 Imola Individual (prototype).

Kardon stereo, electric rear pop-out windows and front-side airbags were also added – one can assume there were enough parts left over from the E36 model run.

For external styling, BMW UK went again for clear indicators with the GT Class II spoiler, front bumper corner extensions – a design cue we would see again in the E46 models – sat on a fresh set of polished forged M Double-Spoke alloys in traditional staggered set-up (7.5J front and 8.5J rear).

The bright interiors weren't for everyone, but sat under the dealer lights the stunning Imola paintwork and its aggressive stance were as close as most would get to the GT race car.

Left: The E36 M3 Imola Individual would pair Red Nappa leather with Anthracite as with the M3 GT Individual before it.

Below left: E36 M3 Imola Individual rear GTII bi-lane spoiler.

Below right: Polished, forged M Double-Spoke wrapped in a Michelin Pilot Sport.

5

The E46 BMW M3

When BMW launched the E46 BMW M3 to the press in November 1997, it did so without the M3 in its line-up. Just over two years later, the M3 was unveiled to the world. There was much anticipation around the launch of the new M3 as the previous model had been such a departure from the original format with its subtle modifications from the base model.

We needed not to worry however as the first images of the M3 emerged in 2000. It was obvious that BMW's M Division returned to the winning formula adopted with the first generation, with its flared wings and arches, and power bulge in the lightweight aluminium bonnet.

For many it was considered a return to BMW's winning recipe: low-weight construction paired with a high-revving motor, be it still in six-cylinder form. Weight was on the up

The 2001 BMW E46 M3 Coupe in Carbon Black.

again to 1,570 kgs with a full tank of fuel and no optional extras, and as with the previous model, the overall size of the car had grown, mainly to accommodate the larger 2,731-mm wheelbase extended by 21 mm from the E36 M3. The track was also increased with the front now sitting at 1,508 mm and the rear sitting at 1,525 mm, up by 86 mm front and 87 mm at the rear from the previous square set-up in the E36 M3 Evolution.

Height had also slightly increased to 1,372 mm from 1,335 mm and length increased to 4,492 mm from 4,433 mm. Width increased from 1,710 mm to 1,780 mm, as in the previous model. As with the E36 M3 before, power would also see an increase, now up to 343 hp with 269 lb-ft, and a red line now at 8,000 rpm thanks to a new iteration of the S50B32, the S54, a red line that is now variable thanks to the oil temperature sensor altering the amber and red warning lights as the car comes up to temperature, lights that double up as shift warning lights on the SMG.

All that power would be transferred to the tarmac through the combination of the new limited slip differential and the six-speed Getrag gearbox available in manual and SMG II, managed by new engine speed-sensitive, variable-assist power steering and M calibrated (stiffened shocks, springs and anti-roll bars; modified rear axle geometry) independent suspension with aluminium double-pivot strut-type front and four-link integral railing-link assembly at the rear, both complemented by a set of Sachs sport shocks.

The S54, which would go on to be the last naturally aspirated inline-6 to be fitted to an M3, was based on the UK version of the BMW S50B32. It again featured a cast-iron cylinder block with an increased bore to 87 mm, a forged balanced crankshaft with larger stroke now at 91 mm. Further upgrades include reinforced forged con rods and new high-compression pistons, which again could be cooled by a set of oil spray nozzles. The head would receive new cast-iron hollow camshafts (with a duration of 260/260, and lift of 12/12 mm) which help decrease the weight of the cylinder head.

As the engine is still using the VANOS system, the same issue can plague the car as with previous iterations if the system has not been regularly serviced, and sadly with the E46 M3 the issues don't stop there. Head gasket failures were common warranty items, and later on in their lives high-mileage cars can suffer from rod bearing wear and water pump failures. And if you thought all this would be a drain on your wallet, it's nothing

E46 M3 dash with its variable red line.

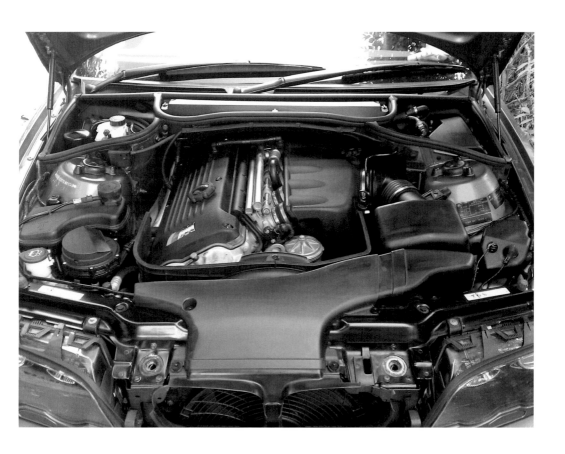

Above: The E46 M3 powerhouse – S50B32 DHOC Twin VANOS 3.2 engine with 343 hp.

Right: The dreaded E46 M3 subframe cracks. (Philip Hopley)

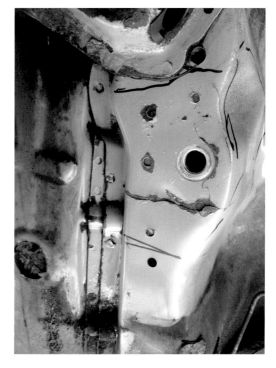

compared to the car's main issue, which is the subframe! A quick internet search will have you running for the hills when considering an E46 M3 on the used car market.

Early models started to exhibit cracking where their rear subframe mounted to the chassis. BMW quickly addressed this issue initially by filling the chassis with an epoxy resin and in cases where the damage had spread, they replaced the actual floor as well as adding the preventative epoxy too. As there was an underlying engineering flaw in the design of mounting points, cars that were fixed could still exhibit cracking in the future.

This opened up a whole market in after-market subframe reinforcement solutions. It's no easy task to repair, as the rear differential and subframe have to be removed to access the damage that requires attention, including some hard-to-reach areas in the boot that require holes cut in the floor, then reinforcement plates are fitted and the car is reassembled. Most cars you will see up for sale will either have this problem or will have had it addressed, and it is something that's very important to consider when entering into a second-hand purchase. The cost of the job can run into the thousands, so factor this into your classified search.

As with the E36 M3 before, the E46 M3 is a well-engineered car, so wears its miles well, subframe issues aside, though the front wings are a common failure point as they rust from the inside out, often caused by dirt and moisture being trapped between the inside of the front wing and the plastic arch protector as they become very inflexible with age. The rear arches are susceptible to rust, as with the previous model, if the arches are not kept clear of dirt build-up, and the door seals still suffer the same issues as the E36 M3 but they are considerably more expensive, and again the exterior trim, such as rear quarter window seals and weather strips, wear out and can be purchased from BMW, though the prices are always creeping up, a side effect of them holding onto new old stock longer.

The rear springs fatigue and crack, so they should be checked, and the auto-dimming rear-view and wing mirrors are prone to failing as they delaminate and can also leak the

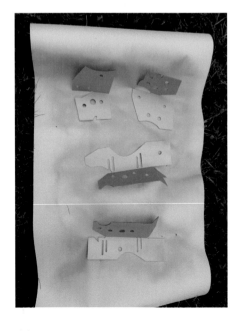

Redish Motorsport E46 M3 reinforcing plates covered in weld-through primer. (Philip Hopley)

LCD liquid, but again there are companies offering fixes or cheap replacements. Finally, the rear diff can be noisy if the latest oil with friction modifier additive hasn't been used, but again easily resolved with a quick oil change using the appropriate oil.

The E46 M3 had presence; if you were behind one or saw one in your rear-view mirror there was no question that it meant business. This time they were unmistakably different to the standard coupe – with an aggressive front bumper, flared front and rear arches, M vent grilles in the front wings, a rear diffuser wrapping quad exhaust tips and the new aluminium bonnet with 'Power Dome' to finish it off.

Wheels were also on the increase, as the standard M-style alloy wheels with Satin Chrome finish and safe-stop rim design had increased to 18 × 8.0J on the front and 18 × 9.0J

Above: E46 M3 SMG in Individual Estoril Blue.

Below: E46 M3 Convertible in Laguna Seca Blue. (BMW Group Global)

at the rear, with asymmetric high-performance tyres: 225/45/18 front and 255/40/18 rear. If they weren't to your taste, there was the very popular Double-Spoke (style 67) polished and forged alloy wheels, again with safe-stop rim design, in a larger 19 × 8.0J front, 19 × 9.5J rear set-up with 225/40/19 front and 255/35/19 rear performance tyres.

There would be no four-door in the E46 model line-up this time; however, a convertible was still present, but not as popular as the coupe, as when you lose the roof you lose the rigidity of the chassis, which goes against the grain in the ultimate driving machine.

There were new colours added to the range, such as Phoenix Yellow and Laguna Seca Blue, which it must be said were very bold choices, especially when combined with matching leather. Now twenty years on, these cars have become quite rare and are often sought after, one such example being Silverstone Blue – there were around fifty units produced in the UK, so it has become a special car in the eye of M3 enthusiasts owing to its limited numbers.

Other colours featured on previous models were included, such as Imola Red, and should you order through the BMW Individual system, you could have any colour you desired, such as Estoril Blue or your favourite colour from the 1970s – making for more unique vehicles amongst the second-hand market. Cars ordered through the Individual program can be identified by the notation on the strut tower and may have even had Individual badging if the owner so wished.

E46 M3 Individual markings.

Step up in luxury with the E46 M3 heated leather interior.

It was clear that BMW was now marketing the M3 as the 3 Series flagship, as specifications were getting greater and greater: air conditioning, electric-heated seats, widescreen sat nav, Harman Kardon sound packages with twelve upgraded speakers (including two subwoofers), upgraded amplification, vehicle-speed-sensitive equalization, as well as various safety features that would come as standard, such as pillar and side airbags added to the driver and passenger pair. Dynamic Stability Control was now a standard option, which fortunately could be turned off, along with a clever new M Variable Slip differential. No longer was the M3 just the framework for a race car, but BMW's prestige does-it-all sports saloon.

New driver comfort and convenience options would be added to the E46 M3, such as a new three-spoke leather-wrapped multi-function M Sport steering wheel with fingertip cruise, audio and (accessory) phone controls, M colour stitching and tilt/telescopic adjustment, rain-sensing windscreen wipers with automatic headlight control and heated side-view mirrors, windshield washer jets and door locks. The passenger's side-view mirror could be set to automatically tilt down when the vehicle is shifted into reverse gear. Park distance control would come as an option, as well as a widescreen TV.

If you opted for the 'Cold Weather Package' you would get three-stage heated front seats, a high-pressure liquid headlight cleaning system and a ski bag, whereas the Premium Package would see a power glass two-way moonroof with remote one-touch operation and

E46 M3 SMG II DriveLogic and all the options ticked.

sliding interior sunshade, power-adjustable front seats with a three-setting memory system for driver's seat and outside mirrors, and Nappa leather interior.

Speaking of seats, the M Vader seats were gone, replaced with larger BMW M front sport seats with adjustable thigh and lumbar support plus the option of electrically adjustable bolster supports.

Not only did the new M front seats have three memory settings, but your seating and mirror memory could now be stored on your key, which made it interesting should you pick up a different key and have the car's seats and mirrors spring into life adjusting for a phantom driver.

Visibility at night was improved by the introduction of Xenon headlights, which were automatically levelled, as the start of BMW's intelligent lighting systems, and the all-new automatic dimming rear-view and side mirrors.

There was no longer a spare tyre, and in its place was the M Mobility System, an air compressor and a bottle of BMW tyre glue, but you did now have a new flat tyre warning system which, although a bit sensitive, was no bad thing as visually it can be difficult to see a drop in air pressure on the wide low-profile tyres.

SMG would feature as an option, as with the E36 M3, though in SMG II format it featured the new Drivelogic system, with the ability to change the ferociousness of gear change as well as rev-matching on the downshifts. Sadly again with the first SMG, the technology was still unreliable and was very expensive to fix when it failed, so consequently it's now becoming more common to see people converting the SMG II box to manual by adding the missing detent springs or replacing it altogether with the manual alternative, which is relatively easy to do as BMW used the same manual gearbox for the SMG II and left space

48

on the pedal set to slide on a clutch pedal – it's almost like they anticipated the conversion. Then it's just a case of someone with the know-how to reprogram the ECU to allow the car to understand it is no longer having its gears changed by a robot.

Fly-by-wire technology was now incorporated into the E46 M3, meaning there was no longer a direct link between you and the throttle linkage. It was not all bad news though, as with the addition of this technology came the sport button, part of BMW's M Dynamic Driving Control, which quickened up the throttle response, making this change more than bearable.

The E46 M3 would see an upgrade to its brakes with a new high-performance-tuned anti-lock braking system (ABS) with Dynamic Brake Control, four-wheel BMW M vacuum-assisted ventilated disc brakes and floating front callipers and larger floating front discs (325 mm). Stopping power was more than adequate for sensible daily road driving, though they were again not really suitable for track use, and are often replaced by big brake kits or at least a set of racing brake pads.

The lack of a decent set of brakes that are track ready wouldn't generally be an issue for us average Joes, running to work and back with the occasional Sunday B road blast. However, it would seem that from the moment the world saw the M3 for the first time it quickly became motor journalists' car of choice for power sliding around racetracks, and would go on to be the performance yardstick to which all other sports saloons were measured and was often seen competing against its closest rivals such as the Audi S4 and then Mercedes C63 AMG – invariably it always came out top with its 50/50 weight distribution, high power to weight ratio and one of possibly the best chassis that BMW ever built.

The E46 M3 production would run from 2000 to 2006, selling 85,766 units, up from 71,242 for the previous model, with a split of 56,133 coupes and 29,663 convertibles, which showed how the popularity of the badge was increasing.

Common upgrade for the E46 M3 – AP Racing big brake kit with six piston front callipers.

6

The E46 BMW M3 CSL

When it came time to produce a special edition in 2003, the BMW M Division decided to release a lightweight special, the BMW E46 M3 CSL. This would be only the second time that the CSL (Competition, Sport, Lightweight) designation had been applied to a car since the original E9 3.0 CSL back in the 1970s.

Weight reduction was not the only element to be addressed with the M3 CSL, but handling and power too, with an eye on breaking the eight-minute lap time on the Nürburgring, as was becoming the performance yardstick of the time.

The newly designed E46 M3 CSL front air dam feeding air to the carbon airbox.

Above: 2003 E46 CSL in the Sapphire Black paused on *EVO* magazine's 'EVO Triangle'.

Right: E46 M3 CSL carbon fibre roof designed to lower the CSL's centre of gravity.

Light weighting was achieved with the clever use of carbon fibre-reinforced plastic in the front and rear bumpers and rear diffuser, along with a redesigned boot lid formed out of lightweight SMC (Sheet Moulding Compound). The inclusion of a carbon fibre roof panel not only helped with the car's diet but also had the effect of lowering the centre of gravity. The net result of all the changes made the car 110 kgs lighter than the standard M3.

From the newly designed front air dam feeding air to the carbon airbox under the aluminium bonnet, with carbon corner flaps and the integrated 'duck tail' rear spoiler, it was clear BMW had been busy at work in the wind tunnels, working hard on reducing lift forces.

All these changes made the M3 CSL quite something, and those in the know could pick it out of a line-up, but more importantly, the M3 CSL had another trick up its sleeve, which was its new, extremely large, airbox and air collector finished in carbon fibre that not only allowed the engine to inhale more air than usual, but produced one of the most addictive sounds known to a petrol head: induction roar, and lots of it, the soundtrack of motorsport for sure.

SMC (Sheet Moulding Compound) integrated rear spoiler on the E46 M3 CSL.

To help the outgoing flow of exhaust emissions, the M3 CSL came with optimised exhaust valves, a modified exhaust emission manifold and funnel-shaped pipes leading into the air silencer. With the addition of higher lift cams (288/280 with 12.5 mm of valve lift) and the new airbox, the output was raised to 360 bhp at 7,900 rpm with torque increased to 273 lb-ft at 4,900 rpm, and these enhancements and power raised the 0-100 km/h dash to 4.9 seconds.

E46 M3 CSL engine bay dominated by the extremely large carbon airbox and air collector.

Though the M3 CSL would only be produced with an SMG II gearbox, the Drivelogic software would be specifically redesigned and readjusted, as well as receiving launch control! As with the previous version of SMG, you could drive as an automatic or sequentially change gears like you were in a touring car without the need to lift off the gas, and there was also a hidden sixth Drivelogic mode, only accessible if you were to turn off the traction control – so you know that the M Division engineers are encouraging sideways fun...

With a wider front track and modified suspension geometry, a new standard of dynamic driving performance was given to the M3 CSL. Michelin would also provide some special rubber to go with the newly devolved 19-inch aluminium alloys 8 ½ J × 19 (front) and 9 ½ J × 19 (rear), of course now fondly known as 'CSL Alloys', in the form of their Pilot Cup Sports, wider than the standard offing at 235/35 ZR 19 at the front and 265/30. Developed specifically for the CSL, they would feature an asymmetric tread and near racing-slick levels of grip, so much so that buying a car with these tyres would require you to sign a waiver that you understood the car would not grip in the wet!

Turning to the interior, you were met with more signs that this was a motorsport special, with lightweight glass-fibre plastic bucket seats, trimmed in Amaretta, as was the steering wheel, featuring only a small 'I/O' button, carbon fibre door cards, centre console with that all-important numbered plaque and two individually contoured lightweight seats in the rear.

Weight was further reduced by having manually adjustable seats, no AC or radio in the base model, and a thinner rear glass window.

Coming back to that little button, this activated M Track Mode, a special DSC mode, and to quote BMW, 'This special function of Dynamic Stability Control (DSC) conceived especially for motorsport allows the driver of the M3 CSL to achieve the highest conceivable standard of longitudinal and lateral acceleration on the racetrack.' Translation: big skids and sideways fun until the car's traction control thinks you are going to have a big accident.

A final nod to the exclusivity. If the limited numbers of 422 UK examples were not enough to show that you had splashed out for the ultimate driving machine, the limited colour choice of Silver Grey Metallic and Black Sapphire Metallic should suffice as evidence.

Above left: E46 M3 CSL DriveLogic program selector with the extra CSL mode.

Above right: The now classic E46 M3 'CSL' alloys, 235/35/19 at the front and 265/30/19 at the rear.

E46 M3 CSL
glass-fibre-plastic
manually adjustable
lightweight sports
seats with Amaretta
and fabric trim.

Later in 2005, as the E46 M3 CSL was coming to the end of its life, there would be another, though perhaps less desirable, special edition: the M3 CS (Clubsport), based on a regular M3 with the addition of CSL wheels, CSL steering wheel and the all-important CSL M differential. The cars were as a standard M3 externally unless you opted for the Clubsport rear spoiler.

No nonsense E46 M3 CSL dash with Amaretta-wrapped steering wheel with the special I/O button.

7

The E46 BMW M3 Went Racing

When it was time to get the E46 M3 competition ready you would think BMW had it in the bag with the high-powered precision-handling M3. Throw on some racing stickers, sit it on some wide racing slicks and get some huge arches to cover them, and off we go. But no, BMW had other ideas and went on to produce one of the most limited production cars of all time: the BMW E46 M3 GTR.

E46 M3 GTR advertising poster for the road-going version of the GTR race car. (BMW Group Classic)

Created in early 2001, the GTR would be the first M3 in history to have a V8 engine, and a 4.0-litre one at that, producing over 480 hp, harking back to BMWs 'go fast' roots – let's take a small mid-sized saloon and slap the biggest engine in there we can find!

Homologation dictated that a road-going vehicle would need to be produced, and BMW set about building ten examples to satisfy the rules. The M3 GTR was born, with one thing in mind: to go racing in the American Le Mans Series under the GT class.

The M3 GTR road car would see a standard M3 get a stiffened chassis, and a sports suspension set-up derived from the race car, along with some additional suspension supports under the bonnet, tying in the firewall to the struct tops. The front and rear bumpers were redesigned as well as extended, and the M3 would receive a new rear wing as part of its aero pack.

Carbon fibre-reinforced plastic would feature heavily to aid with the light weighting effort, with a CFRP roof, rear spoiler, and front and rear bumpers as with the GTR race car.

The interior would be largely untouched, though updated with an eye on weight reduction, so the heavy front and rear seats were removed, replaced with a set of leather-clad Recaros, and rounding out the special interior was a set of special GTR sill plates.

The most interesting element of the M3 GTR would be found under the lightweight bonnet: the high-revving 4.0-litre V8, producing over 350 hp. It would be the most

BMW M3 GTR in Laguna Seca 2001, American Le Mans Series (ALMS). (BMW Group Global)

powerful M3 BMW had so far produced for the road. As it was a detuned P60 racing engine it would feature a dry sump, again another first for an M3 road car. Traction was managed by the use of the twin disc clutch set-up mated to the six-speed manual, and the final drive would see the use of a variable locking M rear differential.

All in all, it was something quite special, and should you have had a spare 250,000 euros, you could treat yourself to a piece of M power history.

In its racing form, the car had great success in its first year, claiming seven out of the ten races it entered, beating the Porsche GT3-R by some margin. However, the following season there were changes in the rules that would result in manufacturers having to produce 100 cars and 1,000 engines to race with no penalties, and though BMW could have continued, albeit with the associated penalties, they decided to pull out of the series, essentially finishing the GTR's short-lived career.

This wasn't the end of the story though, as in 2003 Schnitzer Motorsport would enter two GTRs in the Nürburgring twenty-four hours endurance race, again in the GT class, and in 2004 and again in 2005 it would claim a 1–2 podium finish. The car would also race in the Spa 24 Hours and would see various private teams fit 3,997 cc BMW V8 engines into the E46 M3 and go racing on the Nürburgring, winning some VLN (the Nürburgring Endurance Series) races in later years.

The GTR's success, though brief, was still massively in the shadow of its spiritual ancestor the E30 M3 DTM, and with the M Division's experience, racing knowledge and technological enhancements, one would think the latest racing M3 would conquer all – but it wasn't to be.

8

The E92, E90 and E93 BMW M3

The public waited with bated breath after the launch of the new E9x 3 Series in 2004, as again there was initially no M3 within its ranks. And a wait they had too, until 2007 when the fourth generation of M3 was released. It was designated E92, E93 and E90 respectively, which aligned with the coupe, convertible, the return of the ever-practical four-door.

It was clear that the M3 was growing up; the E9x model was clearly aimed at the discerning executive with its luxurious cabin, adorned with leather everywhere the eye could see, and now adequate room in the back for two reasonably sized passengers, even in the coupe. There were no big spoilers or extended air dams, but subtly contoured side skirts, enhanced by the flared front and rear arches. The front wings, as with the E46 M3,

2007 E92 M3 Coupe in Jerez Black with option 19-inch Satin Chrome finish light-alloy wheels.

Above: E92 M3
Four-door saloon with
its two-door counterpart.

Right: Integrated
indicator with M3 logo
on the E92 M3.

would again get vent gills and would now have integrated indicators and the M3 logo, M rear bumper with diffuser and M front bumper that was modified just enough to let you know that this was not a standard car.

Additional design cues would be carried over from the E46 M3, in the form of the now ever-present power bulge, M quad polished stainless steel exhaust tips, and the most striking of these would be the new carbon roof to feature on the coupe, as seen previously on the E46 M3 CSL. Back were M exterior side-view mirrors in body colour with double bridge-mounting arms as a nod to the E36 M3.

Carbon Fibre Reinforced Plastic Roof (CFRP), part of BMW's M3 continued light weighting program.

As the M3 matured, now in its fourth generation, we started again to see proportions increasing, largely to make room for the luxurious interior and large electrically operated leather sports seats. Weight was on the up, now at 1,655 kgs, an increase from the previous model. Overall dimensions would also increase, seeing the E92 M3 grow 196 mm wider to 1,976 mm, with a square track set-up that would see the front and rear track now sitting at 1,537 mm, up from the E46 M3 before. Finally, the length was increased a further 126 mm to 4,618 mm, accommodating the improved longer 2,761-mm wheelbase. Even height was raised by 40 inches to 1,412 mm, giving a bit more headroom to its occupants.

The M Division had a solution though for all this midlife spread: they had dropped the S54 in favour of a lightweight 4.0 V8, derived from the E60 M5's screaming S85 V10 that had been released just a few years earlier that delivered a bewildering 414 hp and developed 295 lb-ft of torque, and let us not forget this is a family-derived saloon car, more powerful than a 997 Porsche Carrera S of the same year, and with its 8,400 rpm red line it would drown out the flat-six bark in an induction roar sound-off too.

This new engine was something of a celebrity and went on to be a five-time International Engine of the Year winner, as well as ultimately becoming the last normally aspirated engine to power an M3.

The S65, in the E9x M3, would share the same 92-mm bore and 75.2-mm stroke cylinder dimensions as well as the 12.0:1 compression ratio with the fire-breathing V10 it was based on.

Though a larger power plant, the new V8 was in fact lighter than its 3.2-litre counterpart, weighing in at 202 kgs, 15 kgs less than the previous M3s. S54 straight-six, thanks to the use of eutectic aluminium and silicon alloy in the block, and feature a cross-plane, forged-steel crankshaft, magnesium-steel conrods, and iron-jacketed aluminium pistons.

E92 M3 S65 V8 engine.

The heads were constructed of hydro-aluminium and contained two aggressively profiled, double chain-driven cams per bank, once again controlled by a double VANOS system for improved power delivery, better fuel economy and reduced emissions, as with the previous M3 models.

As well as eight individual throttle bodies, the M3 would feature a new high-performance Siemens ECU that was capable of more than 200 million calculations per second. This is rocket scientist levels of computation, in a production car no less!

To further reduce weight and improve reliability, the V8 engine would receive a new lubrication system comprising of a single sump with two electrically operated scavenging pumps.

The technology didn't stop there; it would be the first time an M3 would see a brake energy regeneration system, not to save the planet – as was the case with the EVs made around the same time – but created to reduce the alternator's charging rate during acceleration, to enable greater power delivery.

Sadly, early engines would suffer rod bearing failures because of the rod bearing clearance to journal ratio being smaller than recommended, leading to oil starvation, which in turn causes premature wear. In mid-2010 BMW moved from the original copper and lead bearings to a new improved harder-wearing comping featuring tin and aluminium, as well as increasing the bearing clearance slightly to aid with oil lubrication.

As with the E46, the fourth generation M3s would use a fly-by-wire throttle system, and as the V8 had two throttle banks, it followed that it had two throttle actuators. This would turn out to be another weak point, but once resolved, under warranty or with one of the many after-market offerings, the car would be a very reliable second-hand proposition.

The first E9x M3s are now over fifteen years old, so electrical gremlins are starting to creep into some of the vehicles, as they rely heavily on electronics, even to read the oil level or should the control check system find something it doesn't like, it can even stop you from starting the car.

Maintenance would change on the fourth generation M3, meaning gone were the traditional Oil Service, and Inspection one or Two services, replaced by the new BMW Maintenance Program based on the criteria set by the BMW **Condition Based Service** or **CBS** system, which tells the driver and the dealership what it needs, such as Oil, Service or brake pads, etc. This can make it difficult when checking the service book for a car's history. Ensure the stamps are backed but with receipts, so you can ensure the correct elements have been addressed and the service hasn't been a box-ticking exercise. The first running-in service is, as with the E46 M3, an important receipt to have.

When buying an M car, it's common that they can change hands frequently as people like to tick them off the 'cars to own' list, but a high number of owners and cars changing hands regularly can affect resale prices, and as there are lots of M3s for sale on the used car market, selecting a car that's got a good colour, low owners, low miles, high spec and detailed service history will be a better proposition once you have finished with your V8 fun and come to move on.

The E9x M3 would see the use of the M Drive System, as first seen in the E60 M5, which allowed the driver to store all their favourite M settings, such as steering and throttle response, and the amount of traction control. This could be recalled via the M button on the steering wheel.

You could also now opt for BMW's new clever EDC (Electronic Damper Control), allowing the driver to choose between comfort, normal or sport – yet another adjustment that can be added to the M Drive profile should you so desire.

We would see the word 'Drivelogic' again, but this time gone was the single-clutched SMG gearboxes, replaced with the new M dual-clutch DCT. Taking advantage of modern software and precision mechanics, the DCT gearbox featured seven gears and eleven electronically controlled drive programs. Six of these were intended for manual mode via the shift paddles. Another feature that would find its way from the SMG II was Launch control for maximum acceleration from a standing start. Would you be brave enough to give it a go?

The Drivelogic programs were yet another option that could be configured and saved to the M Drive profile, so you could turn the car from day commuter to weekend track monster at the touch of a button.

In the latest iteration, the new gearbox would allow quicker gear changes than ever before and transmit power to the tarmac faster than had been seen in any previous M3. The DCT-equipped M3s paired with the variable M differential would hit the 100 kph barrier in a claimed 4.5 seconds. For the advocates of pure driving, the manual six-speed transmission was of course still available should they want to do the same dash but take slightly longer doing it, as the manual was a claimed 0.2 seconds slower.

Power is nothing without traction, and this was provided with a larger M Double-Spoke (Style 219M) Satin Chrome finish light-alloy wheels, 18 × 8.5J at the front and 18 × 9.5J to the rear shod with 245/40/18 front and 265/40/18 rear performance tyres or optional Double-Spoke (Style 220M) polished and forged light-alloy wheels, 19 × 8.5J front, 19 × 9.5J rear with 245/35/19 front and 265/35/19 rear performance tyres. Either way, the grip

The E92 M3's new dual-clutch, seven-speed DCT gearbox with DriveLogic.

was well managed thanks to the revised M suspension incorporating aluminium and lightweight steel construction, paired with M Sport gas-pressurized front struts and rear shock absorbers.

Braking was provided by four-wheel BMW M vacuum-assisted ventilated, cross-drilled compound disc brakes with floating brake rotors, now paired with a high-performance-tuned anti-lock braking system (ABS) with Dynamic Brake Control.

Moving to the interior, upon arrival you are greeted by the familiar BMW M3 logo door sills and luxurious welcoming Novillo leather power-adjustable front sport seats with four-way lumbar support, manually adjustable thigh support, heated (if that is your thing), with a memory system for driver's seat and outside mirrors, as with the outgoing model before. Added to the E92 M3 were a new Seatback easy-entry feature with memory and 2x-speed power fore-aft control and new extended leather options that would see leather on virtually every surface you could see.

A further sign that the M3 had its eye on dominating the executive market was the long list of options you could now tick off, with a keyless access system introduced to the M3 for the first time, as well as adaptive headlights with high-beam assist that would detect an oncoming car and dip the main beams for you, and even help light around the corner if needed.

E92 M3 Novillo leather power-adjustable front sport seats with four-way lumbar support.

TV would see an upgrade to digital, as would radio with DAB, and you could choose to add the 12 GB hard drive to your entertainment system allowing you to transfer your old CD library to the car as well as your personal mp3 music collection thanks to the iPod® and USB adaptor ports. You could also now opt to hook up your phone to a Bluetooth hands-free system should you wish to hold a conversation whilst enjoying the V8 soundtrack.

Connected service offerings would start to be used across the BMW range, and the M3 would get its share, with access to BMW's mobile internet portal including services such as news and weather. There was also a clever SOS button should you find yourself broken down or in need of assistance, combined with BMW's automatic vehicle location service.

More options 'packages' were now offered to give your new M the edge, with the Technology and Premium packages being added to the Cold Weather Package as before.

Most notable was the Technology pack, which combined M Drive with M Dynamic Mode (MDM) for accessing and designating M-specific driving systems, electronic Damping Control with Comfort, Normal and Sport modes, new comfort Access keyless entry with multi-function remote control, comfort opening plus a navigation system with 16:9 widescreen and voice-controlled iDrive system that would include real-time traffic information and six programmable memory keys.

Though you still had to push a key into a slot in the dash to wake the car, you now had an engine start/stop button to wake the V8. The Check Control vehicle monitoring system was still keeping an eye on all things electrical, as well as now measuring your oil among other things, which is all fine and well; however, if one of the sensors was to suggest to the system that you didn't have enough oil, the nanny system would stop you starting the car, and this is the same for any of the core systems, such as the brakes.

Auto start/stop would be also introduced to the E92 M3, to complement the Brake Energy Regeneration and lightweight engineering principles that were already firmly established with the new car. This would be the first time BMW would use the technology on a model with more than four cylinders and a transmission other than a six-speed manual.

After the E92 M3 was established in the marketplace, BMW would offer a new Competition Package as an upgrade pack that could be added to your new car's options list. Aimed at sharpening up the driving dynamics, at an additional cost of £3,315 the Competition Package added reprogramed EDC dampers receiving a new sport setting that featured new damping rates, and it would sit 10 mm lower than the standard car. To let your neighbours know you had spent the extra cash, the Competition Package would bring with it a nice set of 19 CSL-style alloys.

Around this time BMW introduced their Frozen paint colours, the first of which being Frozen Grey. This colour was created by applying a special matte clear coat over a Space Grey Metallic base coat. The matte effect was quite something on an E92 M3 Coupe, as it defined the lines of the car as it played with the light, reflecting around 20 per cent of the light that hits it, as well as adding another level of individuality.

Frozen paint, though visually impressive, did come with some drawbacks, as you must not polish a matte finish, and car care is limited to regular shampoo and insect remover. BMW would recommend theirs at this point; no other detailing products are needed. In fact, BMW warned in their Individual Frozen paint car and maintenance guide, 'You can even damage Frozen matte paint by using a microfibre cloth, was, sealant or quick

E92 M3 Coupe in Frozen Grey matte finish – insects beware...

detailer', so if polishing your pride and joy and smothering it in wax and glaze on a Sunday afternoon is your thing, Frozen paint is perhaps not for you.

The main issue with a matte finish is paint correction, as if damaged, even with a slight scratch, the only option for remediation is to repaint the panel, and ask any body shop, matte finish is not an easy finish to pull off, as the application is a one-time deal as of course you cannot polish out any mistakes. Further complications arise by not knowing how many coats to apply and at what thickness to get the right contrast, as blending the paint is not an option.

Perhaps, that's why when BMW continued to use the matte-effect finishes, they would create the Frozen effect using a matte effect clear wrap rather than creating all the magic in the paint alone?

The F30 is a relatively new car compared to the previous models at the time of writing, but it has been around long enough for the internet to start to fill up with common issues. It's wise to look out for overheating radiators and engine modifications.

There are reported issues with the timing, excessive oil usage, gearboxes randomly selecting neutral, and water condensation getting into the very expensive sealed LED headlamps. This is further confirmed by owners complaining about interior rattles and failing mounts and bushes. As the miles are eaten up on British roads, the wear and tear will start to take its toll, so always look out for a good maintenance record, and try and take a prospective purchase on a decent test drive over various road conditions.

The E92 M3 Would Hit the Track

The E92 M3 would become the track experience car of choice for MSV when they launched their M3 Track Master experience, where you could drive an E92 M3 Coupe around one of their racetracks with a qualified race driver on hand for some instruction, and then be treated to a hot lap in the hands of one of their not-so-tame racing drivers. Here you would soon learn about the car's real capabilities. These cars were largely just collected from the dealer and had some vinyl stickers slapped on them before being thrashed by all and sundry for 10,000 miles, and they did it without breaking a sweat, such is the proficiency of the E92 M3.

It would be twenty years since the works team exited DTM and returned, and they would do so in 2012 with the E92 M3. As with the E30, they would gain early success with Bruno Spengler securing the top step of the podium on the car's second ever race, in his Black BMW Bank M3 DTM. No one would have wished for a better return to the DTM after such a long time away, but that would only be the start of it, as after four more race wins, and a nail-biting final race against Gary Paffett at Hockenheim, Spengler would become the DTM champion that year, and to round out what was already an overwhelming success, BMW and BMW Team Schnitzer would also collect the Manufacturer and Team titles, mirroring that of the E30 M3's early DTM success.

The M3 DTM in E92 trim would only race for two seasons, replaced by the M4 DTM, but in those two seasons BMW would collect another Manufacturers title for the 2013 season and Augusto Farfus would finish runner-up – not a bad return to the DTM all things considered.

Above: A line-up of MSV E92 M3 Coupes at Oulton Park.

Below: Nürburgring (2010) BMW M3. Team BMW Motorsport, No. 25, Jörg Müller (DE), Augusto Farfus (BR), Uwe Alzen (DE), Pedro Lamy (PT), No. 25. (BMW Group Global)

9

The E92, E90 and E93 BMW M3 Special Editions

M3 Frozen Silver Edition

When it came to model specials, the E92 would see them come thick and fast. The first of these was the BMW M3 Coupe Frozen Silver Edition, with a limited run of 100 units. It featured as standard the seven-speed M DCT transmission and variable M differential and was lowered by 10 mm from standard with three-stage Electronic Damper Control, 19-inch light-alloy wheels and a revised DSC+ system.

E92 M3 Coupe in Frozen Silver Edition. (BMW Group UK)

For the first time we would see Frozen Silver metallic paint paired with matte black alloys, dark chrome exhaust tips, side grilles and kidney surrounds, and jet black air intakes on the bonnet. A high-gloss Shadow line trim would complete the exterior look.

The standard Black Novillo leather interior trim was complemented with contrast stitching and Palladium leather armrests and inserts, and the steering wheel, trimmed in Alcantara, came complete with an M3 chequered flag motif, which was mirrored on the door sills.

The Frozen Silver Edition would receive its special edition plaque as well as just the right amount of carbon fibre trim to highlight the car's sporting pedigree.

Above: E92 M3 Coupe in Frozen Silver Edition interior with Palladium leather inserts. (BMW Group UK, Riess FotoDesign)

Right: E92 M3 Coupe in Frozen Silver limited edition plaque. (BMW Group UK, Jamie Cottington)

M3 GTS Coupe

Next we would see the closest thing to rival the E46 BMW M3 CSL in the E92 format: the 2010 E92 BMW M3 GTS Coupe. Again a limited run of 150 units would be produced.

It was instantly recognisable wearing its orange paint finish, large adjustable rear wing, extended front air dam and black-style 359 alloy wheels, 19x9.0J front and 19x10.0J rears, wrapped in 255/35/19 front and 285/30/19 performance tyres. A high-gloss Shadow line trim would again feature on the exterior, matching the side vents and bonnet intake ducts.

As with the E46 CSL that came before, lighter materials were used in the door panels and polycarbonate was chosen for the rear windows. Gone too were the executive leather seats, replaced by a set of lightweight Recaro seats and a colour-coded roll cage. All this compromise would save around 50 kgs, and those who knew where they were going and enjoyed the soundtrack of a high-revving V8 could also save more weight choosing the free option of dropping the sat nav, multimedia system and the automatic air conditioning, another carry over from the E46 M3 CSL for the hardcore racer.

Pop up the bonnet and there are more signs of something a bit special, as you are greeted by more orange paint, this time on the engine cover. Go deeper and you will discover that the engine capacity has been increased to 4.4 litres, taking the power to 450 hp, which would be utilised by the ever-present seven-speed M DCT transmission and variable M differential.

Along with the sports seats and roll cage, the E92 M3 GTS Coupe would receive a coil-over suspension set-up, more evidence that BMW had designed this limited edition car to be a serious proposition for the track.

The 2010 E92 M3 GTS Coupe in Fire Orange. (BMW Group Global, Barry Hayden)

Above: E92 M3 GTS Coupe lightweight Recaro seats and its colour-coded roll cage. (BMW Group Global, Barry Hayden)

Below: E92 M3 GTS Coupe powerplant bored out to 4.4 litres, delivering an impressive 450 hp. (BMW Group Global, Barry Hayden)

E92 M3 GTS seven-speed M DCT transmission and Alcantara-covered steering wheel. (BMW Group Global, Barry Hayden)

With all the weight saving and new higher-output motor, the M3 GTS would complete the dash to 62 mph (100 kph) in 4.4 seconds and would go on to reach an impressive maximum speed of just shy of 190 miles per hour.

M3 CRT

In 2011 we would see the BMW M3 CRT, a high-powered four-door saloon with the same power as the GTS it followed, featuring a 4.4-litre V8 again with 450 hp. As the name would suggest, the CRT (Carbon Racing Technology) would see carbon-reinforced plastic components used in its production to dramatically reduce weight, including the bonnet, rear spoiler, seats and air-channelling element integrated into its front air dam, saving 70 kgs over the standard production vehicle.

Only sixty-seven vehicles would be produced, finished exclusively in Silver Frost with some subtle red accents around the bonnet air ducts, side grilles and front air dam.

Unlike the GTS, comfort and practicality had not been sacrificed, and looking at the interior, though the CRT received CFRP fixed-back bucket seats and an Alcantara-covered steering wheel, it also retained creature comforts such as DCT with Drivelogic, Professional sat nav, a BMW Individual High-End Audio System, a light exterior mirror and luggage area package, an alarm system and front and rear park distance control.

Above: The 2011 E92 M3 CRT finished in Silver Frost and red accents. (BMW Group Global)

Below: E92 M3 CRT CFRP fixed-back two-tone bucket seats and Alcantara-covered steering. (BMW Group Global)

Occupants were reminded they were in something unique by the limited edition plaque featured on the dash, as well as by the roar of the new lightweight flap-controlled sports exhaust with titanium mufflers, which was further enhanced by sound-proofing configured specifically for the new model.

As with the M3 GTS, the M3 CRT would also receive an adjustable coil-over suspension and upgraded brakes, rounding out the whole package, adding the CRT to BMW's long line of exclusive lightweight sports cars that can trace its lineage back to the 1970s BMW 3.0 CSL, as had the E46 CSL before.

2012 M3 DTM Champion Edition

If it was exclusivity you were after, you couldn't go wrong with the DTM Champion Edition. Rarer than the M3 CRT, the M3 E92 DTM Champion Edition would be a limited run of just fifty-four cars corresponding to their 54 DTM wins, as well as the Driver's title, won by Bruno Spengler.

In 2012, after a nearly twenty-year absence, BMW would return to the DTM and in doing so they exceeded all expectations, with their driver Bruno Spengler winning the Driver's title. BMW Team Schnitzer would take the Team honours, and the cherry on the cake? BMW claimed the Manufacturer's title too.

In doing so, BMW Motorsport achieved a total of fifty-four DTM race wins, and that is exactly the number of limited edition models that would go on to be produced.

The cars would be finished in Frozen Black matte, with carbon fibre flaps, and a gurney. The cars were based on the look of Spengler's racing car, so would receive sections of the race car's livery along with some tasteful dark chrome embellishments.

The race design would be continued over the exterior with a BMW M stripe design over the carbon-reinforced plastic roof and boot lid, and feature the BMW M logo on the wings and decals in the rear-side windows.

2012 M3 DTM Champion Edition in Frozen Matte Black. (BMW Group Global)

Above: 2012 M3 DTM interior dash would be quite special featuring Spengler's driver's signature and the car's serial number. (BMW Group Global)

Below: 2012 M3 DTM Palladium accents amongst the extended black Novillo leather trim. (BMW Group Global)

It was not everybody's cup of tea, but makes a fitting tribute to the success of the Canadian driver. Tributes were continued into the interior, in the form of unique door sill strips that were designed to replicate his helmet design, and the carbon fibre interior dash strip would display the driver's signature and the car's serial number.

The additions didn't stop there. The steering wheel would be trimmed with Alcantara, as with earlier special editions, and again there would be some Palladium accents amongst the extended black Novillo leather trim and an M power-lettered embroidered handbrake lever.

Although visually the car looked like a race car, it would still receive the M level of refinement that had come to be the hallmark of BMW's flagship model – essential daily driver equipment such as the Professional navigation system, heated seats and Park Distance Control. And to ensure the car could still be driven like a race car, 'Technical' elements like the Competition Package, M Drive, M DCT Drivelogic and the M Driver's Package were ready and waiting.

The M3 LE 500

The UK would again then receive yet another special edition offering in the form of the BMW M3 Coupe and Convertible Limited Edition 500, produced exclusively for the UK and patriotically offered in Imola Red, Mineral White or Santorini Blue.

Exterior differences would again see the M car's signature details, such as kidney grille surround, side grilles and twin double exhaust tips finished in dark chrome, whilst BMW's

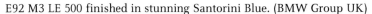

E92 M3 LE 500 finished in stunning Santorini Blue. (BMW Group UK)

high-gloss Shadow line exterior trim would be used to further accent the exterior. 19-inch black alloy wheels would complete the look.

As the name would suggest, the run would be limited to 500 models (coupe and convertible combined), and each one would feature an edition designation laser cut into the trim inlay on the dashboard.

The Limited Edition 500 was based on the standard M3 platform and so had the standard high-revving 420-hp V8 engine with M-specific chassis, with either a six-speed manual or seven-speed M DCT transmission.

M3 Coupé Edition Model

As was becoming a common theme, BMW released yet another E92 M3 Coupe Edition model. This time the focus would be on colours available to UK customers: Alpine White, BMW Individual Dakar Yellow and BMW Individual Monte Carlo Blue. Edition models would feature dark chrome shadow-coloured kidney grille, side vents, bonnet vents and twin exhaust tailpipe, as seen before on previous special editions, but as an extra piece flair, they would receive black wing mirrors with connecting supports in the car's body colour, as a sign that this was a standard M3.

M3 Coupe Editions feature 19-inch M Double-Spoke light-alloy wheels finished in silver alloy as standard and with a black gloss finish as an option. High-gloss Shadow line exterior trim would feature externally as standard.

E92 M3 Dakar Edition at Oulton Park Circuit.

E92 M3 sports seats with body-coloured stripes on the standard Novillo leather interior, with colour-coded contrasting stitching.

On the inside there would be colour-coded contrasting stitching on the leather trim and body-coloured stripes on the standard Novillo leather interior, on the front and rear seat backrests and on the extension of the seat squab. If you opted for Dakar Yellow or Monte Carlo Blue, the sticking was continued on the armrest and passenger door handle.

The Alpine White Edition M3 had an interior with contrast stitching on the carbon structure leather trim and contrast-coloured door armrest pads. The centre console and centre armrest pad were finished in Alpine White, which is not a great colour for that area. Door sill plates would feature a checkered flag to underscore the M3's racing pedigree of the M3, reminding those who enter of the M3's racing heritage.

M Performance Edition (2012)

One of the final Specials to be released before the fourth generation finished its production run was the M Performance, limited to thirty units and available in M3 Coupe format.

Each car would be identified by a laser-cut 'One of 30' label on the dash cover. Mechanically the cars were identical to that of the M3 Competition Package, with a 414-hp 4.2-litre V8, M DCT paddle shift transmission, and rear-wheel drive.

For the extra bucks you would have to stump up, you received £9,790 worth of optional equipment that came as standard, including a Harman Kardon thirteen-loudspeaker system, BMW Professional Media Package, DAB radio and those clever adaptive headlights,

Right: E92 M3
M Performance
in Frozen Red
with carbon fibre
corner splitters.
(BMW Group
Global)

Below: E92 M3
M Performance
Alcantara
and piano
wood interior
and M Sport
multi-function
seats. (BMW
Group Global)

along with some Alcantara and piano wood trim, sliding front armrest and some heated M Sport multi-function seats.

Those BMW M Performance Edition colours look rather special too, sat on a set of 19-inch M Y-spoke Matte Black alloy wheels, complemented by Dark Chrome exhaust tips, side gills and kidney surround, jet black bonnet intakes, carbon fibre front splitters, and of course the now classic high-gloss Shadow line exterior trim.

Frozen paint options would again feature, and the M3 was available in Frozen Blue, Frozen White or Frozen Red. Black trim on the kidney grilles and wheels completed the M performance's look.

Production of the M3 Convertible would continue for a bit longer than the coupe and saloon, but by October 2013, E9x production would be completed, with 40,092 coupes, 16,219 convertibles and 9,674 four-door saloons produced.

E92 M3 M Performance 19-inch M Y-spoke matte black alloy wheels and high-gloss Shadow line exterior trim against Frozen Blue. (BMW Group Global)

10

The F80 BMW M3 and F82 BMW M4

2014 would see the fifth generation M3 be released and again there would be a four-door saloon. But hang on, there is no other option, as the coupe and convertible were released as an M4! BMW had started their new marketing model, splitting its compact executive model's offerings into two different series designations with the more family-orientated cars keeping the traditional 3-Series badge, having four doors, and the sportier-looking vehicles like the coupe and convertibles gaining a new designation, the M4.

What was BMW doing? Had they just disregarded twenty-five years of racing heritage? The M3, the most successful touring car of all time, the two-door coupe, was now a four-door family car, and its replacement, the M4, no longer wore the iconic M3 badge.

Saddened though the diehard M power fans were, this wouldn't be the only thing to set the forums on fire. With an eye to the environment, the E92's screaming V8 would make way for a smaller six-cylinder unit, a 3.0-litre twin-turbocharged unit, the S55, producing

Fifth generation F80 M3 with its two-door replacement, the F82 M4 Coupe. (BMW Group Global, Daniel Kraus)

425 hp and 550 Nm of torque. With a claimed 100 km/h in just 4.3 seconds with the optional automatic dual-clutch gearbox, it was clear that the changes in engine philosophy were not just for the planet.

BMW are not the only manufacturer opting for smaller engines that are boosted by turbochargers, and if it is good enough for a McLaren supercar, then it's good enough for a four-door sports saloon.

The F80 M3 would receive a new colour pallet, and bold metallic options such as Yas Marina Blue, Austin Yellow and Sakhir Orange would become new enthusiast favourites.

The size was again on the way up, though one would expect an increase as let us not forget this is a family car! Overall length and height had grown to 4,686 mm and 1,425 mm respectively, making the new F80 68 mm longer than the E92, and a modest 13 mm higher, with more changes in width and wheelbase. The front axle would now sit at 1,577 mm wide and the rear at 1,605 mm; the wheelbase would be further increased by an additional 51 mm to 2,812 mm. The wider wheelbase would house the new 18-inch forged light-alloy wheels, 18x9J front and 18x10J rear, with 225/40/18 performance tyres at the front, and 275/40/18 at the rear, or you could choose to fill that wider wheelbase with a set of 19-inch option wheels, 19x9J front and 19x10J with a 255/35/19 front and 275/35/19 performance tyre combo.

Though now a four-door, it didn't mean the BMW M3 was having a midlife crisis; in fact, it was the opposite as BMW's lightweight design concepts were used in the car's production and would see a weight saving of around 80 kgs over the outgoing M3, showing BMW's focus was still on making a racing car for the road, and to that end BMW took what

F80 M3 3.0-litre twin-turbocharged S55 M power engine. (BMW Group Global)

they had learned from their involvement in motorsport technology, running thousands of laps of the legendary Nürburgring Nordschleife – the world's most challenging, exacting racetrack – to dial in the cars for that transition.

The M Division engineers worked tirelessly on the development of the new models, underpinning the creation of two high-performance sports cars that BMW claims 'set new standards in terms of the overall concept, precision and agility'.

The new high-revving six cylinders in the latest M cars featuring M Twin Power Turbo technology newly developed for the new BMW M3 Saloon and new BMW M4 Coupe produces a maximum power of 431 hp, with a peak torque of 406 lb/f. This is available across a wide rev band; BMW claim that it outstrips the figure recorded by the outgoing BMW M3 by roughly 40 per cent, but the engine also achieves a reduction in fuel consumption and emissions of around 25 per cent. This would be the first time BMW would use turbos on their M cars, but would not be the last...

The Twin Turbo M3s were capable of a claimed 0 to 60 mph in 3.9 seconds (with the seven-speed M Double Clutch Transmission 4.1 with the Manual), which is again faster than the outgoing model.

Moving to the inside, the interior was very familiar, coming largely from the standard 3 Series, though that's no bad thing – there is no doubt though that you are in the M, with a liberal covering of M stitching and carbon fibre accents, new M Sport heated leather seats that hold you firmly in place with their contoured sides, raised elements further enhancing the overall design.

The M leather steering wheel would again be multi-functional, with the all-important M Drive button to store your car's personalised set-up. DCT models would receive metal-finished gear shift paddles too.

The M logo dash would now feature a new information panel, showing the driver their chosen drive settings, such as suspension, throttle and steering, and a heads-up display

Above left: F80 M3 vent grille with integrated indicator and M3 logo. (BMW Group Global)

Above right: Quad tip exhaust system on the F80 M3 and subtle boot lip spoiler. (BMW Group UK, Stuart Collins)

Left: F80 M3 M Sport's heated leather seats with raised elements and the M logo. (BMW Group Global)

Below: M Technology on the F80 M3 including the new iDrive data display sat proudly on top of the new dash. (BMW Group UK, Stuart Collins)

Aerodynamically optimised Twin Stalk M mirrors, a design carried over from an outgoing model. (BMW Group Global, Tom Kirkpatrick)

would now feature on the M3, having previously debuted on the S60 M5, but now with more information – speed, navigation direction and check control data now available to help you keep your eyes on the road.

Technology would continue to be applied to the M cars, as well as the latest iteration of BMW's connected service, where new driver-assist programs are designed to monitor everything around the driver, including traffic, road signs, pedestrians, and animals, as well as keeping an eye on the current environmental conditions and alerting you of any potential hazards.

Externally too there were the usual M enhancements to the standard model, including the M power bulge bonnet now synonymous with the M3, as were the wider flared arches and wings with a deep front spoiler now with three large intakes under the newly designed trademark BMW kidney grilles, giving the new family-friendly F80 M3 a very aggressive look. The lightweight carbon roof would be carried over from the outgoing model but would now be aerodynamically profiled and the vent grilles in the front wings would again have an integrated indicator and the M3 logo.

Twin Stalk M mirrors are used as a design cue from previous generations, aerodynamically optimised with new bespoke LED indicators.

Lightweight design is a high priority in the construction of all BMW M cars, and the F80 M3 would receive its fair share of ultra-lightweight materials including magnesium, aluminium, and a firm M favourite, Carbon Fibre Reinforced Plastic (CFRP), as well as new carbon fibre prop-shaft.

When it comes to getting the car on the road, all these further weight reductions enhance the drivability and handling of the new F80 M3, and the grip is managed by the latest iteration of the electronically controlled Active M differential. Steering inputs are now made using the new M-developed electric power-assisted steering with Servotronic that automatically adjusts the steering sensitivity based on the car's speed. Whether navigating a tight spot in the multistorey car park, piloting a narrow and twisting B road or negotiating a high-speed run on the Autobahn, the Servotronic system ensures maximum agility and comfort.

F80 M3 in its natural habitat. (BMW Group Global, Daniel Kraus)

Braking was finally addressed with the F80 M3, and six-piston fixed callipers would be introduced with the new M compound brakes. The two-pieced drilled form discs would now sit at 380 mm, connected again to a central aluminium hub. The rears would feature a traditional floating single-piston design but carry the same-designed 370-mm disc.

If high-speed runs are your thing, you'll be glad to know BMW now also offered a Carbon-Ceramic option with 400-mm discs up front and 380 mm at the rear. To let your mates down the pub know that you were a more discerning customer, the callipers would be finished in gold paint, reserved exclusively for the Ceramic Package.

Cooling was improved too, thanks to the three air ducts in the M air dam, now optimised and widened to deliver the oncoming air straight to the brakes.

EDC would again feature with the F80 M3's Adaptive M Suspension, with its three electronically controlled driving modes – Comfort, Sport and Sport+ – though no matter which option you select, the F80 M3 chassis is firm – as you would expect from an M car.

The F80 would be produced between 2014 and 2020, and during its run, as with the previous model, it would receive a Competition Package, introduced between 2016 and 2018.

The F80 M3 Competition Package

The special Competition Package from BMW M GmbH would combine added sporty personality with enhanced dynamics. As well as a host of handling upgrades, the

Competition Package came with additional equipment and an increase in power that would see a new max output of 444 hp from the 3.0 Twin Turbo M Motor.

Extra power meant improved performance, and the new BMW M3 with the optional seven-speed M Double Clutch Transmission (M-DCT) sprinted from 0 to 60 mph in just 3.8 seconds (without Competition Package: 3.9 sec).

As with the E92 M3 before, the Competition Package would include Adaptive M Suspension, which has been extensively tuned to enhance performance and handling, with new springs, dampers and anti-roll bars, along with reconfigured driving modes (Comfort, Sport and Sport+). The standard Active M differential on the rear axle and DSC (Dynamic Stability Control) would be reconfigured to match the upgraded dynamics.

No M3 CSL-style options this time, but a set of multi-spoke, machine-polished, weight- and rigidity-optimised, forged 20-inch M alloy wheels (front 20x9J and 20x10J rear) with a set of performance tyres (front: 265/30/20, rear: 285/30/20).

F80 M3 Competition multi-spoke, machine-polished, forged 20-inch M alloy wheels with exclusive gold callipers signify the Ceramic Package had been chosen. (BMW Group Global)

M3 Competition special lightweight M sports seats. (BMW Group Global)

Weight- and rigidity-optimised forged, machine-polished, multi-spoke 20-inch M alloy wheels (front 20x9J and 20x10J rear) with mixed tyre sizes (front: 265/30/20, rear: 285/30/20).

The interior would receive a set of special lightweight M Sport seats designed to add additional support on the track and offer exceptional comfort on the street. Seatbelts with woven-in BMW M stripes finish off the sporty look of the cabin.

Externally, the Competition Package brings with it an M Sport exhaust system with black chrome tailpipes. This not only looks great but adds a visceral resonance to the driving experience, from the rumble at start-up to the burbling overrun.

Extended BMW Individual high-gloss Shadow line exterior trim would again feature to tie in with the already standard high-gloss Black finish on the side window trim, window recess finishers and exterior mirror frames and bases. Again as with previous special editions, the Competition Package would include the high-gloss Black finish on the BMW kidney grille, the side grilles and now for the first time on the model badge on the boot.

11

The F80 BMW M3
UK Special Editions

The M3 CS

In 2018, as the production of the M3 was starting to come to an end, BMW M would release another exclusive: a limited-run special-edition model in the shape of the new BMW M3 CS.

Following in the footsteps of the successful M special editions which began in 1988 with the E30 BMW M3 Evolution, the M3 CS was designed to deliver a high level of everyday practicality and dynamism at the same time, which is the true beauty of an M car, a family race car for the shops.

M3 CS and its carbon roof and state-of-the-art twin LED headlights. (BMW Group Global, Uwe Fischer)

The Twin Turbo engine would receive yet another bump, with an additional 10 hp more than with the Competition Package, raising it to 454 hp, making it one of the highest-power and fastest M cars BMW's M Division had produced to date, with a 0–62 mph time of 3.9 seconds.

The M Driver's Package would be a standard option on the new special edition, as was the raised electronic speed limiter, set to a conservative 174 mph! Power and performance would come from the M3's twin mono-scroll turbochargers, charge air cooler, High Precision Injection system, Valvetronic variable valve timing and the double VANOS unit. A new red start/stop button would alert you to the fact things were about to get interesting, as you brought the M3 CS to life.

As with the Competition Package, the BMW M3 CS would get its own unique specially tuned sports exhaust, which would make the hairs on the back of your neck stand to attention.

The lightweight M3 CS would be fitted with the seven-speed M Double Clutch Transmission (M DCT) as standard with Drivelogic, allowing for seamless gear changes to be made in fractions of a second with no lift-off of power, whether running in automatic mode or taking charge yourself with either the gear stick or the paddles on the steering wheel.

BMW M3 CS would receive all the enhancements from the Competition Package that came before, and BMW again looked to reduce all unnecessary weight, with various

Press to awaken the beast – the BWW M3 CS's red start/stop button. (BMW Group Global)

M3 CS rear with its specially tuned sports exhaust, Gurney spoiler and rear diffuser. (BMW Group Global, Uwe Fischer)

suspension components being constructed from extremely light aluminium, ensuring that the car's unsprung weight was kept at a minimum to improve handling response.

The M3 CS is fitted as standard with Adaptive M suspension, whose geometry has been tuned to deliver optimised performance on both the road and track, paired with the Active M differential and Dynamic Stability Control system, which would include the special M Dynamic Mode.

The control systems for the Adaptive M suspension, DSC and Active M differential were also modified to suit the dynamic requirements of the M3 CS, as was the electrically controlled steering.

As has become the fashion with the new M cars, the driver has the option of adjusting the steering and suspension settings to their personal preferences or the demands of their chosen journey.

Another nod to the legendary E46 M3 CSL would see the M3 CS wearing Michelin Pilot Sport Cup 2s (front: 265/35/19, rear: 285/30/20) fitted as standard on its light-alloy wheels, (front: 19x9J, rear: 20x10J), showing that the car would be as much at home on the racetrack as it would be on the M6. The road-legal semi-slick cup tyres provide unparalleled levels of grip and a high level of lateral stability at high track speeds, and as with the E46 M3 CSL, should you not wish to sign a disclaimer, and actually use the car in the rain, you could order up the car with a set of Michelin Sport road tyres.

Moving to the inside of the M3 CS, you are greeted by a two-tone full-leather interior finished in Silverstone/Black and Alcantara, with yet another nod to the E46 M3 CSL in the guise of the Alcantara steering wheel. Brimming with technology, including automatic climate control, a Harman Kardon surround sound system and a Professional sat nav, maintaining the high standard of trim that we have come to expect with BMW's special editions and the M Division has certainly given the interior a look that reflects the car's sporting intent.

On the exterior, the M3 CS would receive lightweight and extremely rigid high-tech materials including a carbon fibre-reinforced plastic (CFRP) bonnet, as well as the signature

Left: M3 CS light-alloy wheels with carbon ceramic brakes. (BMW Group Global)

Below: Two-tone, full-leather interior finished in Silverstone/ Black and Alcantara. (BMW Group Global)

carbon roof. The front apron with large, three-section air intakes, and state-of-the-art twin LED headlights are a prominent feature of the M3 CS front end, and the specially designed M3 CS boot lid would get a new Gurney spoiler in addition to the front splitter and the rear diffuser, helping to minimise dynamic lift.

The M3 wouldn't feature as BMW's race car of choice and was passed over by the rebadged M4, so technically not an M3, though if it walks like a duck and it swims like a duck... The two-door M Coupe of its time was still representing the brand on the racetracks around the globe, and still battling it out with its road-going rivals. The M4 was developed in 2013 and took part in various DTM races between 2014 and 2018.

The BMW 30 Jahre M3

The final flourish of the F80 M3 would be the BMW M3 '30 JAHRE M3', launched in the summer of 2016, this year chosen as it was exactly thirty years after the signing of the first contract for the purchase of a BMW M3 and would serve to pay a special homage to the thirty-year success story of the BMW M3.

BMW M offered up an exclusive special edition that would be limited to 500 cars, and the UK would only see thirty, all finished in BMW Individual Frozen Silver, accentuating the car's muscular form, and they would only be available with the Competition Package.

Additional limited edition touches would include BMW Individual high-gloss Shadow line with enhanced features, including a black chrome tailpipe trim for the M Sport exhaust system as well as a variety of carbon fibre additions, including the front splitter and trims, rear spoiler, rear diffuser and mirror caps.

The M3 30 Jahre and its grandfather, the E30 M3 Sport Evolution. (BMW Group Global)

Above: The UK BMW M3 30 Jahre (30 Years M3) Edition in Individual Frozen Silver. (BMW Group Global)

Left: 30 Jahre M3 carbon fibre front splitter. (BMW Group Global)

The optimised features would include new springs, dampers and stabilisers, reprogrammed Adaptive M suspension (Comfort, Sport and Sport+) as well as a matched reprogrammed standard Active M differential on the rear axle and Dynamic Stability Control feature (DSC).

The 30 Jahre M3 would get the Competition Package's forged 20-inch M light-alloy wheels with 666 M star spoke design and mixed tyres (front: 265/30/20, rear: 285/30/20), the largest set to feature on an M3 to date.

The interior would be full-leather Merino trim in bi-colour Black/Fjord Blue or optionally in Black/Silverstone with colour-matched contrasting seams, complemented with skeletal-designed M sports seats that were both comfortable and supportive. '30 Years

Above: 30 Jahre M3 M Sport full-leather Merino trim in bi-colour Black/Fjord Blue. (BMW Group Global)

Right: 30 Jahre M3 door sill trim. (BMW Group Global)

M3' would be stitched into the front headrests, and the seatbelts would have BMW M strips woven into them.

'30 Years M3 1/500' wording would be present on the carbon fibre interior trim strip of the instrument panel as a sign of the exclusivity on this special model, and the door sills would display the logo '30 Years M3', reminding both the driver and their front passenger of the BMW M company icon's unique history.

As well as the striking BMW Individual Frozen Silver paint finish, the 30 Jahre M3 would feature exclusively designed M gills in the front wings bearing the logo '30 Years M3'. This celebration of all things BMW M3 would have 450 BHP.

At the end of the F80 M3's production run a total of 34,677 M3s would have been sold compared to over 111,000 of the F82 M4, which goes to show the continued popularity of the formula that started it all over thirty-five years ago, which was taking a medium-sized two-door saloon car, making it handle, dropping in a high-performance motor and a couple of sports seats; spin off a homologation special to go racing and let that technology drive the model's evolution.

With manufacturer design ethos being firmly pointed toward an electric future, we wonder how long it will be before the high-revving, octane-devouring soundtrack of an M-powered monster will be a thing of the past. Will the M3 have been replaced or will it evolve into the king of the executive, electric, mid-sized sports saloon?

Left: '30 Years M3 1/500' displayed on the carbon fibre dash panel. (BMW Group Global)

Below: Four generations of BMW M3. (BMW Group Global, Ralph Wagner)